Physics of Nanostructured Solid State Devices

Physics of Nanostructured Solid State Devices

Editor

Sakshi Mathur

scitus
academics

Physics of Nanostructured Solid State Devices

Edited by **Sakshi Mathur**

ISBN: 978-1-68117-227-9

Library of Congress Control Number: 2015936585

SCITUS Academics LLC,
www.scitusacademics.com
Box No. 4766, 616 Corporate Way,
Suite 2, Valley Cottage,
NY 10989

Printed in United States of America on Acid Free Paper ∞

Contents

Preface

Nanostructured materials with dimensions, i.e., are increasingly interested in nanotechnology Size of grain, shapes or thickness of the layer below 100 nm. Nanomaterials should fulfill two conditions, i.e. Nano is one of the dimensions and certain properties of these materials are specific to the nano dimensions. The use of nanostructure materials for optoelectronic devices, including light emitting diodes (solar cells, photodetectors, laser diodes and LEDs), has received considerable attention recently due to their unique geometry.

This book introduces readers to theories and concepts such as semi-classical and quantum mechanical descriptions of electron transport, methods for calculating band structures in solids with optical constant calculations and other advanced concepts. This book presents some recent developments in the theoretical calculation, characterization and synthesis of these novel nanostructures.

Chapter 1

Improving Light Extraction of Organic Light-Emitting Devices by Attaching Nanostructures with Self-Assembled Photonic Crystal Patterns

Kai-Yu Peng and Da-Hua Wei

Graduate Institute of Manufacturing Technology and Department of Mechanical Engineering, National Taipei University of Technology (TAIPEI TECH), Taipei 10608, Taiwan

ABSTRACT

A single-monolayered hexagonal self-assembled photonic crystal (PC) pattern fabricated onto polyethylene terephthalate (PET) films by using simple nanosphere lithography (NSL) method has been demonstrated in this research work. The patterned nanostructures acted as a scattering medium to extract the trapped photons from substrate mode of optical-electronic device for improving the overall external quantum efficiency of the organic light-emitting diodes (OLEDs). With an optimum latex concentration, the distribution of self-assembled polystyrene (PS) nanosphere patterns on PET films can be easily controlled by adjusting the rotation speed of spin-coater. After attaching the PS nanosphere array brightness enhancement film (BEF) sheet as a photonic crystal pattern onto the device, the luminous intensity of OLEDs in the normal viewing direction is 161% higher than the one without any BEF attachment. The electroluminescent (EL) spectrum of OLEDs with PS patterned BEF attachment also showed minor color offset and superior color stabilization characteristics, and thus it possessed the potential applications in all kinds of display technology and solid-state optical-electronic devices.

INTRODUCTION

Organic light-emitting diode is one of the optical-electronic devices for the various applications not only in general illumination but also for consumer displays due to the probable advantages of excellent EL performances, good flexibility, light weight, better durability, low power consumption, and convenient fabrication. Therefore, the OLEDs seem to be one of the potential candidates for the next-generation emissive display technology [1–5]. However, many research articles have reported that almost 70~80% of the photons generated inside the organic layers is confined in the OLEDs caused by three major factors: (1) the total internal reflection (TIR) occurred at the glass/air interface and organic layers/substrate (substrate and waveguiding mode); (2) the reflection loss is due to the refractive index mismatch between interfaces (Fresnel loss); (3) the emitted light is lost due to absorption and plasmonic dissipation of the metal electrode (absorption and SP mode) [6–9]. For the reduction of the waveguiding mode of the organic layers, various methods such as introducing textured surface, inserting low-index materials, and fabricating patterned nanostructures into the device have been proposed [10–13]. But the light extracted from organic layers still suffers from the light confinement of the device substrate. Using light-coupling structures onto the surface of device can help extract light suffering from the TIR at glass substrate/air interface, thus increasing the total external quantum efficiency (EQE) [14, 15]. Many researches have proposed different ways to extract the trapped light in the substrate mode, including rough or textured substrate surface, microlens array (MLA) film, diffusive particles, and silica colloidal gel [14–16]. It is worth mentioning that, due to the limitation of the standard photolithography technique, the used microlens with diameters near the wavelength of the emitted light from the OLEDs cannot be made. In the presented work, we introduced a facile and effective fabrication method named as nanosphere lithography for the production of periodic particle array surfaces with nanometer-scale features. NSL was firstly reported by the research work of Hulteen and Van Duyne [17] and has been widely employed in the area of two- and three-dimensional (2D and 3D) micro/nanostructures patterning, such as quantum dots, nanowires, nanomesh, antireflection structure (ARS), and 3D inverse-opal photonic crystals [18–23]. Compared to those traditional sophisticated lithographic methods, NSL has shown a great many advantages that include a simple nanofabrication technique, an inexpensive route for the patterning of long-range periodic nanostructure arrays in a large scale area, and high throughput. Methods for the deposition of PS nanosphere solution onto desired substrates include electrophoretic deposition, electrochemical deposition, Langmuir-Blodgett- (L-B-) like technique, drop-coating, and spin-coating which we utilized spin-coating in our present research work [24–28].

Here, we demonstrated an unsophisticated way to effectively extract the trapped photons in the glass substrate to the outer media by attaching the PS nanosphere patterned brightness enhancement film (BEF) onto the OLEDs surface. Owing to the large increase of light extraction from trapped photons, the total luminous intensity enhancement for the device with a single-monolayered nanospheres arrangement patterned film attachment in the normal direction shows a great improvement of 161% when compared to the traditional indium tin oxide (ITO) based OLEDs without any BEF sheet attachment. The OLEDs attached with PS patterned BEF also possessed potential application in flat panel consumer display technology due to the minor color offset and good color stabilization properties in different viewing angles [29].

EXPERIMENTAL PROCEDURE

Figure 1 shows the schematic diagrams of the fabrication process for the preparation of a single-monolayered PS array BEF and its application in OLED devices. At first, the methanol solution mixing with a surfactant Triton X-100 (400:1 in volume) is prepared to dilute the purchased

PS beads. Then the surfactant-free nanometer-scale polystyrenes with latex colloids (Bangs Laboratories Inc., ~10 wt% in water, PS mean diameters: ranging from 200 to 800 nm) were further diluted by the above-mentioned methanol solution in ultrasonic bath. Those as-prepared diluted PS nanospheres with latex solution were spun-coated onto clean PET substrates by the spin-coater as shown in Figure 1(a). The distribution of a single-monolayered self-assembled PS nanosphere arrays can be easily controlled by adjusting the rotation speed of spin-coater as shown in Figure 1(b). The period of the hexagonal patterns is determined by the initial diameter of PS beads and attaching the PS nanosphere array BEF sheet as a photonic crystal pattern onto the OLED device as shown in Figure 1(c). The product information and experimental details for the PS nanospheres with the periodicity ranging from 200 to 800 nm are listed in Table 1.

Table 1: The product information and experimental details are for the PS nanosphere with different periodicities

Diameter of PS balls (nm)	**Product number**	**Content of solid PS in stock solution (%)**	**Solution proportion (methanol : PS latex)**	**Concentration of PS beads in prepared solution (%)**	**Spin speed (rpm)**
200	PS02N	10.0	2 : 1	3.33	2000
300		10.0	1 : 1.5	6.00	1400
400		9.7	1 : 1.5	5.82	1100
500	PS03N	10.1	1 : 2	6.73	800
600		10.5	1 : 2	7.00	600
700		9.9	1 : 2	6.60	500
800		10.1	1 : 2	6.73	400

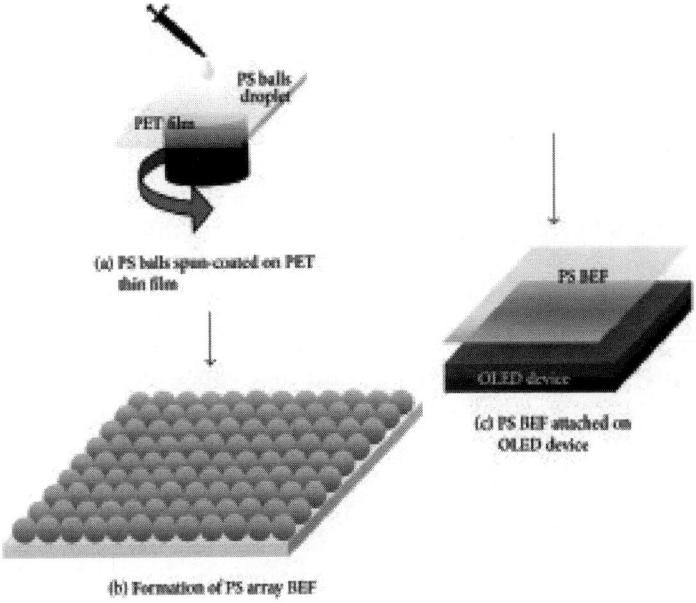

Figure 1: The schematic diagrams of (a) the fabrication procedures for (b) the preparation of single-monolayered self-assembled polystyrenes (PS) arrays, and then (c) the brightness enhancement film (BEF) and its application attached onto organic light-emitting diodes (OLEDs) devices.

The commonly used techniques for fabricating the OLEDs are separated into two major types, including dry and wet-chemical processes. The wet-chemical process includes the spin-coating and inkjet-printing methods, which is suitable for the fabrication of the polymer OLEDs [30, 31]. However, the dry process (vacuum evaporation deposition) applied in our work is mainly used to fabricate the small molecule OLEDs [15]. In the parts of the blue phosphorescent OLED fabrication process, the FIrpic-based light-emitting diode was made onto an ITO coated glass substrate (ITO film thickness: 200 nm, sheet resistance: $10\,\Omega/\square$). Prior to the organic layers evaporation process, the ITO glass was thoroughly cleaned with acetone, isopropyl alcohol, ethanol, and deionized water in sequence and then treated with an UV-ozone cleaner. The multilayers structure for the OLEDs is depicted in Figure 2(a). The 40 nm thick poly (3,4-ethylenedioxythiophene) : poly(4-styrenesulfonate) (PEDOT : PSS) thin films were spun onto the ITO coated glass to reduce the possibility of electrical shorts. The organic layers were continuously deposited onto the substrates in an ultrahigh vacuum (around 5X10^{-6} torr) by thermal evaporation deposition system. The hole-transporting layer (N,N′-diphenyl-N,N′-bis(1-naphthyl)-(1,1′-biphenyl)-4,4′-diamine, α-NPD), blue emissive layer (CDBP:3% FIrpic), electron-transporting, and hole-blocking layer (aluminum (III) bis(2-methyl-8-quninolinato)-4-phenylphenolate, BAlq) were sequentially evaporated onto the substrates. Then LiF and thick Al were set as the opaque cathode for the OLED device. The detailed device structure is composed of ITO (200 nm)/PEDDT:PSS (40 nm)/α-NPD (30 nm)/CDBP:3% FIrpic (40 nm)/BAlq (40 nm)/LiF (1 nm)/Al (150 nm). Figure 2(b) shows the molecule structures of α-NPD, CDBP, FIrpic, and BAlq, respectively. The surface morphology was observed by field emission scanning electron microscope (FESEM, Hitachi S-4800). In order to prepare the SEM specimens, the dilute PS nanosphere latex was spun-coated on the flat PET film thus forming a single-layered close-packed PS array at first. Then the as-prepared samples were put into a vacuum sputtering system to deposit 5 nm Au film. The conductivity of those samples is increased and the point discharge effect would be decreased. Finally, those above-mentioned SEM samples were cut into two pieces for the preparation of the SEM cross-sectional analysis. The EL spectrum, current-voltage characteristics, and luminance for optical-electronic devices were examined by a measuring system consisting of a spectrophotometer (Minolta CS-1000) connecting with a computer and a power supply unit (Keithley model 2400) to drive the OLEDs. The angular dependence of the EL intensity was measured by putting the OLEDs device vertically on the center of the rotation stage.

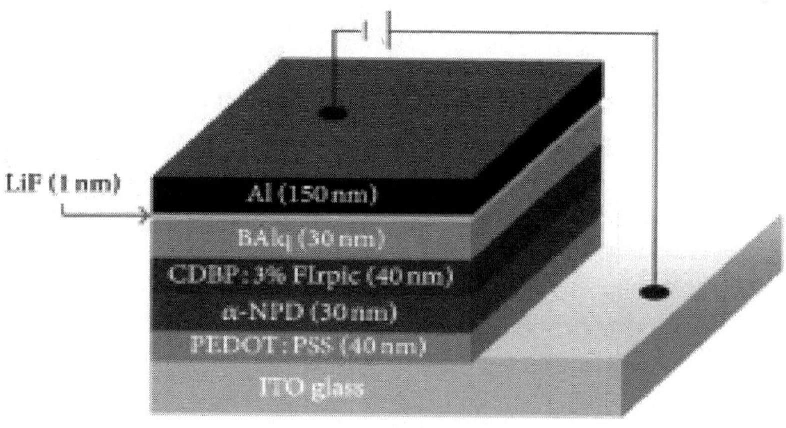

(a)

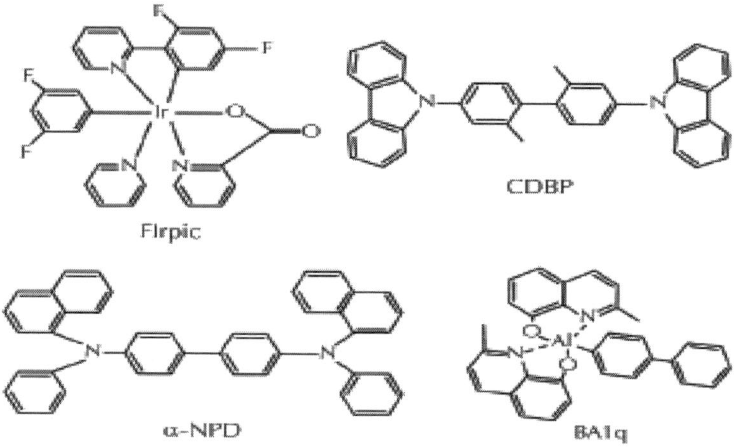

(b)

Figure 2: (a) The diagram is for the OLEDs device structure, and (b) the chemical structures of the molecules employed in the presented research work.

RESULTS AND DISCUSSION

In the typical spin-coating process, the PS balls followed the lowest energy level rule of self-assembly and rearranged onto any kinds of the substrates freely till the solution was dried. Following the lowest energy level rule of self-assembly, a single-monolayered hexagonal closed-packed structure with uniform distribution is formed onto the PET films after few minutes. The period of this hexagonal pattern was determined by the diameter values of PS nanospheres. Since the assembly process occurs during the drying step, two parameters including the colloid latex concentration and rotation speed of spin-coater both play a main role in achieving the monolayer hexagonal and closed-packed PS patterns over the large areas. With a specified latex concentration, the distribution of PS nanosphere patterns on PET films can be easily controlled by simply adjusting the rotation speed. Figures 3(a)–3(c) show the SEM images for 400 nm PS nanospheres spun-coated at different rotation speed conditions after the self-assembly process. The rapid evaporation and stronger centrifugal force occurred at a spin speed of

2000 rpm resulting in the obvious empty space between nanoparticle islands because there is no enough time for those PS nanospheres rearranging into well-ordered 2D islands as shown in Figure 3(a). On the other hand, if the spin speed was reduced to a lower value such as 500 rpm, the slow evaporation rate and weak centrifugal force would lead the consequence into randomly distributed single- or multilayered PS balls stacking as shown in Figure 3(b). Figure 3(c) shows that a single-monolayered close-packed PS pattern formed onto PET sheet in a large area can be achieved by tuning the rotation speed of spin-coater to a proper value, such as 1100 rpm for 400 nm PS nanospheres in the presented work. Figure 3(d) shows the enlarged SEM image of a single-layered close-packed 400 nm period PS nanosphere patterns enlarged from Figure 3(c), and the inset image is the corresponding cross-sectional morphology. Figure 3 also shows the SEM images for (e) 200 and (f) 800 nm PS nanospheres arrays spun-coated at rotation speed of 2000 and 400 rpm, respectively. The distribution results for self-assembled PS nanosphere patterns with 400 nm periodicity at different spin speed parameters are summarized and listed in Table 2.

Table 2: The distribution results for the self-assembled PS nanosphere patterns at different spin speed conditions

Diameter of PS balls (nm)	**Concentration (methanol : PS latex)**	**Spin speed (rpm)**	**Corresponding image**	**Results**
400	1 : 2	2000	Figure 3(a)	Random arranged PS patterns with obvious defects
400	1 : 2	500	Figure 3(b)	Single- or multilayered PS balls stacking
400	1 : 2	1100	Figures 3(c) and 3(d)	Single-layered and close-packed PS patterns

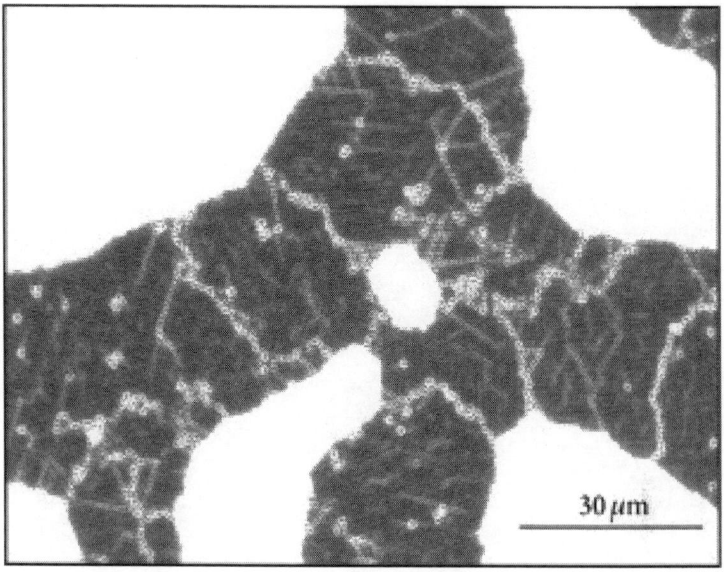

(a)

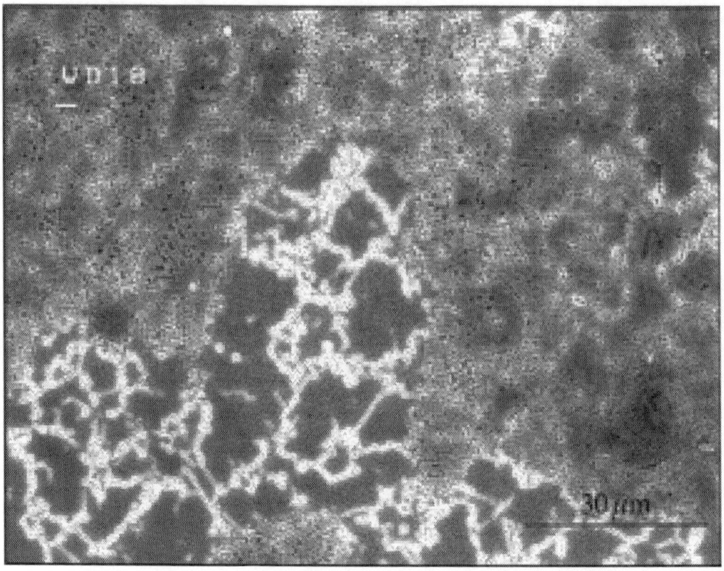

(b)

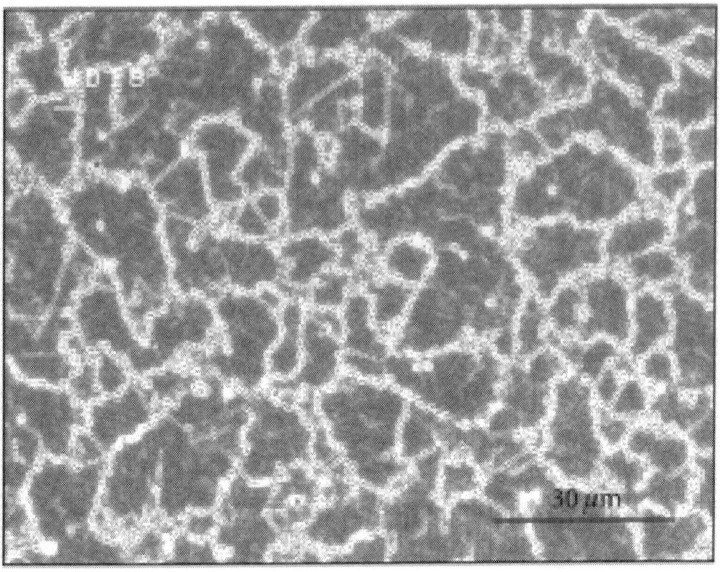

(c)

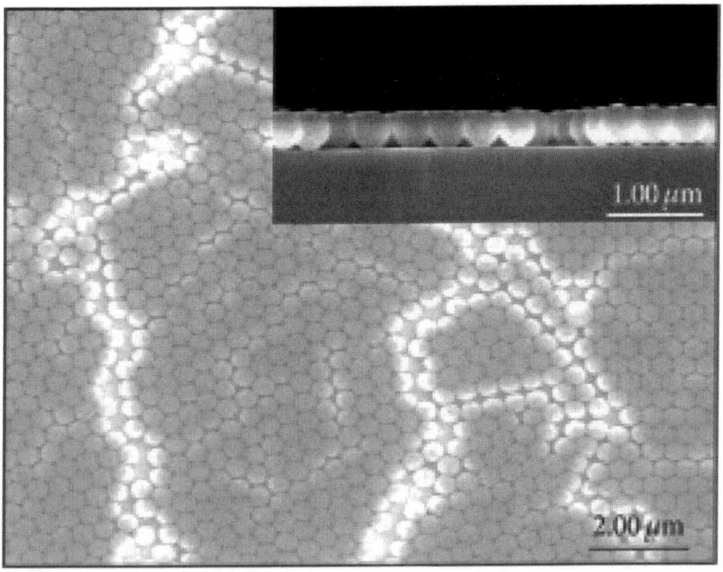

(d)

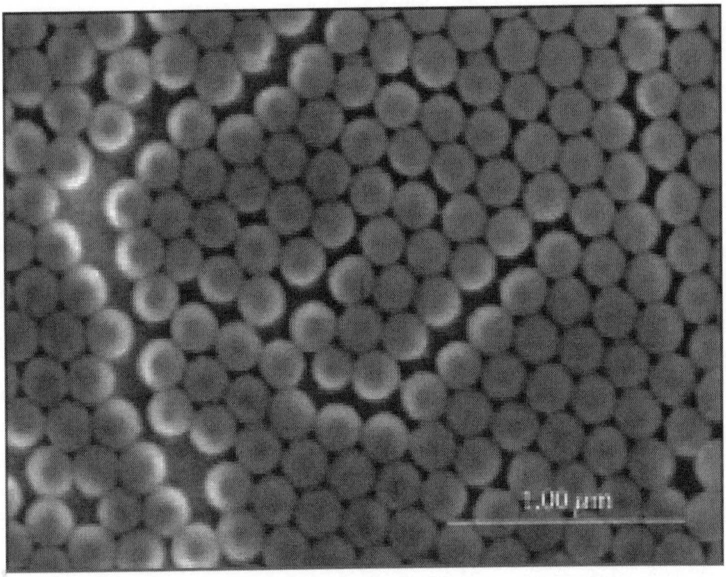

(e)

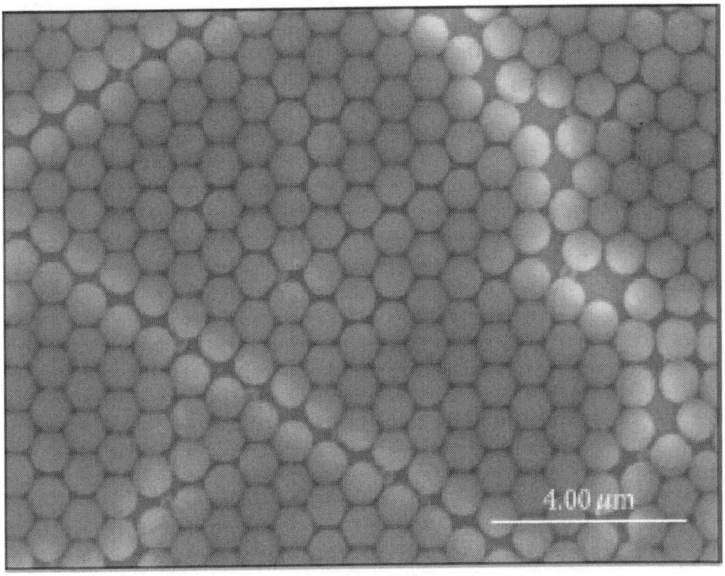

(f)

Figure 3: SEM images are for the 400 nm period PS nanospheres distribution at spin speed of (a) 2000 rpm, (b) 500 rpm, and (c) 1100 rpm, respectively. (d)

The enlarged SEM image is from (c) for a single-monolayered nanosphere array patterns with hexagonal close-packed uniform distribution and the inset image is the corresponding cross-sectional morphology. SEM images are for PS nanospheres arrays with (e) 200 and (f) 800 nm periodicity at spin speed of 2000 and 400 rpm, respectively.

The emission peak for the used OLED device in the presented work is located at the wavelength of 466 nm and with a subpeak at 497 nm as shown in Figure 4(a), which agreed with the work reported by Tokito et al. [32]. The turn-on voltage and the saturated current efficiency of the OLED device at the current density of 10 mA/cm^2 are 8.7 V and 7.0 cd/A, respectively, as shown in Figure 4(b).

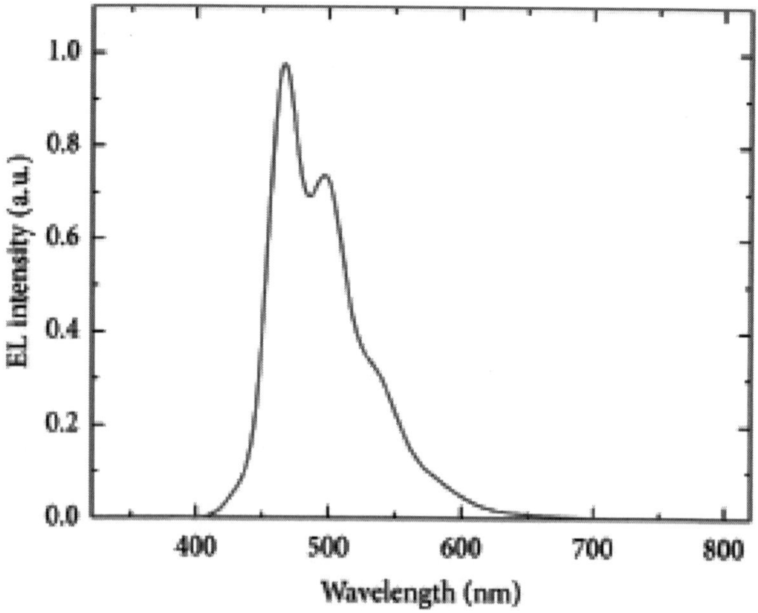

(a)

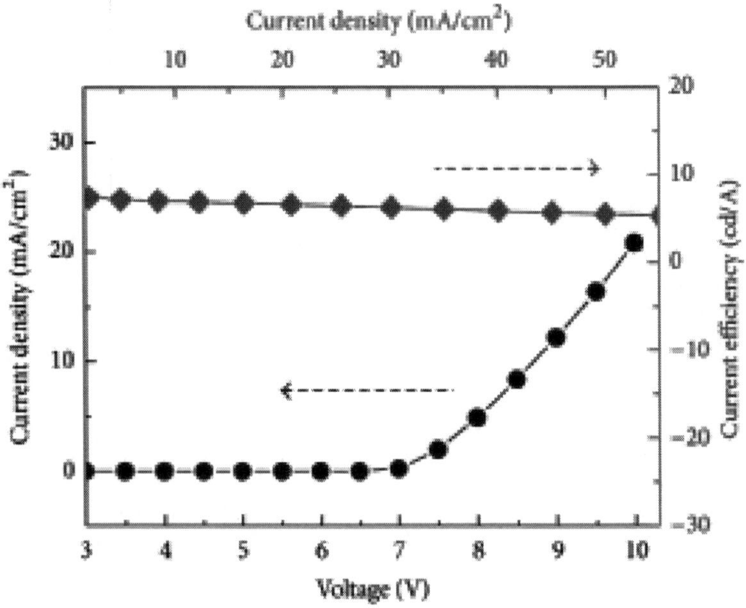

(b)

Figure 4: (a) EL spectrum of the OLED device. (b) J-V characteristic and current efficiency versus current density of the OLED device.

The schematic diagram as depicted in Figure 5(a) illustrated that most of the emitted light generated from the organic layer was confined in the device because of the waveguiding effect and substrate mode. In general, less than one-third of the generated photons can escape from emitting devices in typical OLEDs. One-third of these photons are guided in the glass substrate and the others are trapped in the organic layers. The photons trapped in the glass substrate can be further coupled out when applying a patterned BEF sheet attached onto OLEDs as shown in Figure 5(a). Using the coupling equation can help to understand the coupling of light in the optical-electronic device. The diffraction angle (θ_b) for the light trapped in the substrate is calculated by

$$\frac{2\pi}{\lambda_0} n_b \sin\theta_b \pm m\frac{2\pi}{\Lambda} = k_{\text{glass}} = \frac{2\pi}{\lambda_0} n_b, \tag{1}$$

where m=1 (the diffraction order), k_{glass} is the propagation constant in the glass, n_b=1.52 for a typical BK7 glass, $\lambda_0 = 466$ nm, and Δ =400 nm (the periodicity of the PS nanospheres) in this case. θ_b is 136^0 smaller than the refraction angle in the air (, calculated by Snell's law) demonstrating the trapped photons can escape from glass to air. The insets in Figure 5(b) show the optical images of OLEDs with and without 2D PS nanosphere patterned film attachment. It reveals that the overall luminous intensity of OLED device with 2D PS photonic crystal patterned film attachment is higher than the referenced device. The PS patterned film acted as a BEF to help in extracting trapped photons in the glass/air interface to the outside. The luminous enhancement of OLED devices attached with 2D photonic crystal patterned films of various nanosphere diameters (from 200 to 800 nm) is also shown in Figure 5(b). It can be seen that the luminous enhancement is increased with the decrease of the PS nanosphere diameter values for the photonic crystal patterns with the periodicity of 400–800 nm. According to the diffraction term $(2\pi / \Delta)$ as shown in the grating coupling equation, the PS balls with 400 nm diameter (e.g., $\Delta = 400$ nm) show the greatest luminous enhancement efficiency and the total luminous enhancement for the device with applying PS nanosphere arrangement patterned film attachment shows a great improvement of 61% when compared to the traditional ITO-OLEDs without any BEF attachment owing to the large increase of light extraction from trapped photons. The luminance enhancement factor then drops as the PS diameters in the range between 200 and 300 nm. The results can be concluded into two reasons: (1) the scattering effect of the PS ball pattern is decreased because the diameter values of the PS are smaller than the wavelength of visible light; (2) the diffraction term () is increasing as the PS diameters decrease; therefore, the calculated diffraction angle would be negative, hence lowering the scattering effect caused by the PS nanosphere patterns.

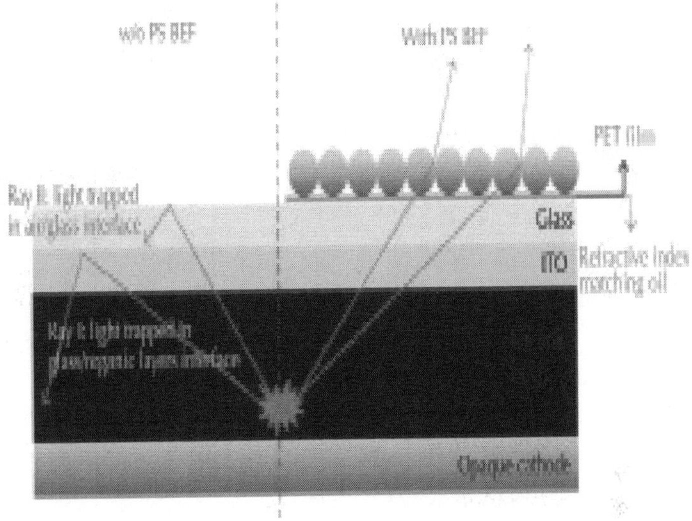

(a)

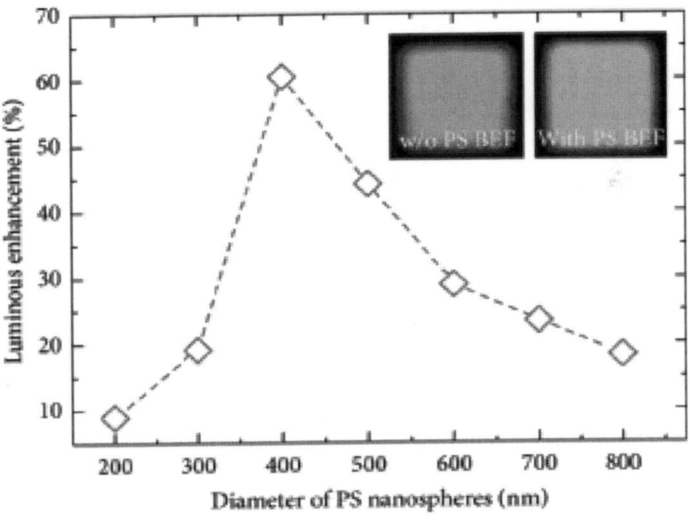

(b)

Figure 5: (a) Light extraction of trapped photons in glass substrate and organic layer of bottom emission OLED and (b) luminous enhancement of OLED devices attached with 2D photonic crystal pattern structures of various nanosphere diameters ranging from 200 to 800 nm.

We further measured the emission light at different viewing angle (VA) to confirm the light extraction enhancement efficiency. Figure 6(a) shows the luminance values for the OLEDs with and without PS nanosphere patterned BEF sheet attachment under 30 mA/cm^2 operating current density in normal viewing angle that are about 3050 and 1900 cd/m^2, respectively. The brightness improvement is about 60.5%, which is in good agreement with the result of luminous enhancement for 400 nm PS patterned BEF sheet attaching onto OLEDs as shown in Figure 5(b). The luminance intensity of device as a function of viewing angle with a constant current injection (60 mA/cm^2) is shown in Figure 6(b). The luminance intensity of device without PS patterned BEF attachment drops as the viewing angle increases. When we applied a patterned BEF sheet onto OLEDs, the angular emission pattern was broadened and the luminance intensity was also increased. Figure 6(c) shows the normalized EL spectrum for OLEDs without attaching PS patterned BEF sheet. The normalized intensity in the wavelength of 497 nm was changed from 0.741 (VA = 0°) to 0.849 (VA = 50°). It revealed an obvious color offset property in the wavelength of 497 nm at different viewing direction. But the color offset phenomenon can be minimized after applying the PS patterned BEF as shown in Figure 6(d). The normalized intensity was slightly shifted from 0.799 (VA = 0°) to 0.771 (VA = 50°). The reason for the minor color offset of the OLEDs attached with PS BEF sheet is attributed to the omnidirectional light harvesting enhancement, which is consistent with the result as shown in Figure 6(b). The emission spectrum also shows that there is almost no wavelength shift when utilizing such a PS patterned BEF attachment, which performs excellent color stabilization characteristics. The results can be confirmed by the Commission Internationale de L'Eclairage coordinates (CIE 1931) and the corresponding CIE coordinates calculated from the EL spectrum for OLEDs without and with PS patterned BEF attachment are (0.1635, 0.2798) and (0.1619, 0.2776), respectively.

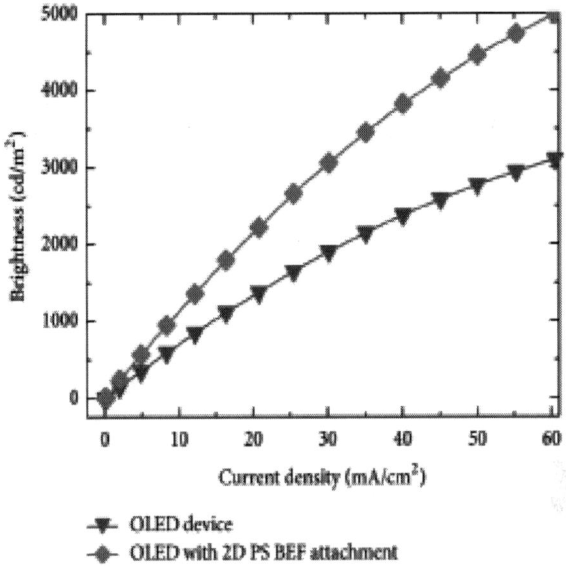
5000
4000
3000
2000
1000
0
Brightness (cd/m²)
0
10
20
30
40
50
60
Current density (mA/cm²)
OLED device
OLED with 2D PS BEF attachment

(a)

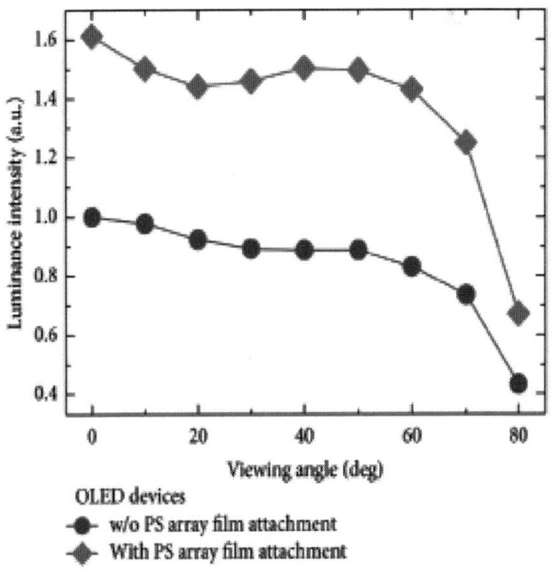
1.6
1.4
1.2
1.0
0.8
0.6
0.4
Luminance intensity (a.u.)
0
20
40
60
80
Viewing angle (deg)
OLED devices
w/o PS array film attachment
With PS array film attachment

(b)

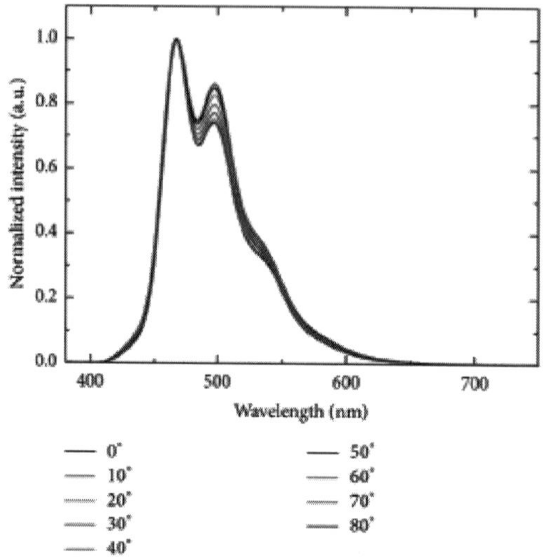

(c)

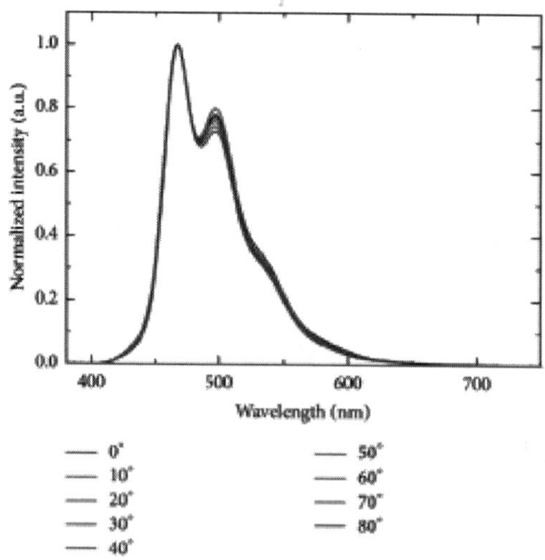

(d)

Figure 6: (a) Luminance versus current density and (b) the luminous intensity as a function of viewing angle for the OLED device with and without PS patterned BEF attachment. The EL spectrum for the OLED device (c) without and (d) with attaching 2D PS array patterned films (diameter: 400 nm).

CONCLUSIONS

In summary, using an external patterned BEF sheet to efficiently improve the total OLED's luminous efficiency has been successfully demonstrated by a facile NSL technique in the presented research work. With a specified latex concentration, the distribution of PS arrays on PET films can be easily controlled by only adjusting the rotation speed of spin-coater. The luminous intensity of OLEDs with attaching PS patterned sheet in the normal viewing direction is 161% higher than the one without any BEF attachment. The presented experimental results indicated that the trapped photons are further coupled out from the substrate mode, thus increasing the light extraction efficiency. The EL spectrum of OLEDs with PS patterned BEF attachment shows minor color offset and outstanding color stabilization property that possess potential future applications in all kinds of optical-electronic display devices and solid-state lighting technology.

ACKNOWLEDGMENTS

The authors acknowledge financial support of the main research projects of the National Science Council of Taiwan under Grant nos. NSC 101-2221-E-027-042 and NSC 101-2622-E-027-003-CC2, respectively.

REFERENCES

1. Y. Luo, L. Wang, Y. Ding, L. Li, and J. Shi, "High light-extracting efficiency for OLED directly fabricated on double-side nanotextured silica substrate," Optics Letters, vol. 38, no. 14, pp. 2394–2396, 2013.
2. J. Huang, G. Li, E. Wu, Q. Xu, and Y. Yang, "Achieving high-efficiency polymer white-light-emitting devices," Advanced Materials, vol. 18, no. 1, pp. 114–117, 2006.

3. C. W. Tang and S. A. Vanslyke, "Organic electroluminescent diodes," Applied Physics Letters, vol. 51, no. 12, pp. 913–915, 1987.
4. S. R. Forrest, D. D. C. Bradley, and M. E. Thompson, "Measuring the efficiency of organic light-emitting devices," Advanced Materials, vol. 15, no. 13, pp. 1043–1048, 2003.
5. B. Niesen and B. P. Rand, "Thin film metal nanocluster light-emitting devices," Advanced Materials, vol. 26, no. 9, pp. 1446–1449, 2014.
6. J. B. Kim, J. H. Lee, C. K. Moon, S. Y. Kim, and J. J. Kim, "Highly enhanced light extraction from surface plasmonic loss minimized organic light-emitting diodes," Advanced Materials, vol. 25, no. 26, pp. 3571–3577, 2013.
7. S. Nowy, B. C. Krummacher, J. Frischeisen, N. A. Reinke, and W. Brütting, "Light extraction and optical loss mechanisms in organic light-emitting diodes: Influence of the emitter quantum efficiency," Journal of Applied Physics, vol. 104, no. 12, Article ID 123109, 2008.
8. R. Meerheim, M. Furno, S. Hofmann, B. Lüssem, and K. Leo, "Quantification of energy loss mechanisms in organic light-emitting diodes," Applied Physics Letters, vol. 97, no. 25, Article ID 253305, 2010.
9. S. Y. Kim and J. J. Kim, "Outcoupling efficiency of organic light emitting diodes and the effect of ITO thickness," Organic Electronics: Physics, Materials, Applications, vol. 11, no. 6, pp. 1010–1015, 2010.
10. I. Schnitzer, E. Yablonovitch, C. Caneau, T. J. Gmitter, and A. Scherer, "30% external quantum efficiency from surface textured, thin-film light-emitting diodes," Applied Physics Letters, vol. 63, no. 16, pp. 2174–2176, 1993.
11. M. Slootsky and S. R. Forrest, "Enhancing waveguided light extraction in organic LEDs using an ultra-low-index grid," Optics Letters, vol. 35, no. 7, pp. 1052–1054, 2010.
12. D. H. Wei, W. H. Liao, and K. Y. Peng, "Light guide of au nanostructures for color-filterness optoelectronic display devices," Journal of Nanoscience and Nanotechnology, vol. 12, no. 2, pp. 1341–1343, 2012.

13. Y. Sun and S. R. Forrest, "Enhanced light out-coupling of organic light-emitting devices using embedded low-index grids," Nature Photonics, vol. 2, no. 8, pp. 483–487, 2008.
14. S. Möller and S. R. Forrest, "Improved light out-coupling in organic light emitting diodes employing ordered microlens arrays," Journal of Applied Physics, vol. 91, no. 5, pp. 3324–3327, 2002.
15. Y. H. Ho, K. Y. Chen, K. Y. Peng, M. C. Tsai, W. C. Tian, and P. K. Wei, "Enhanced light out-coupling of organic light-emitting diode using metallic nanomesh electrodes and microlens array,"Optics Express, vol. 21, no. 7, pp. 8535–8543, 2013.
16. T. Tsutsui, M. Yahiro, H. Yokogawa, K. Kawano, and M. Yokoyama, "Doubling coupling-out efficiency in organic light-emitting devices using a thin silica aerogel layer," Advanced Materials, vol. 13, no. 15, pp. 1149–1152, 2001.
17. J. C. Hulteen and R. P. Van Duyne, "Nanosphere lithography: a materials general fabrication process for periodic particle array surfaces," Journal of Vacuum Science & Technology A, vol. 13, no. 3, pp. 1553–1558, 1995.
18. Y. H. Ho, K. Y. Chen, S. W. Liu, Y. T. Chang, D. W. Huang, and P. K. Wei, "Transparent and conductive metallic electrodes fabricated by using nanosphere lithography," Organic Electronics, vol. 12, no. 6, pp. 961–965, 2011.
19. A. Kosiorek, W. Kandulski, H. Glaczynska, and M. Giersig, "Fabrication of nanoscale rings, dots, and rods by combining shadow nanosphere lithography and annealed polystyrene nanosphere masks,"Small, vol. 1, no. 4, pp. 439–444, 2005.
20. S. Ji, J. Park, and H. Lim, "Improved antireflection properties of moth eye mimicking nanopillars on transparent glass: flat antireflection and color tuning," Nanoscale, vol. 4, no. 15, pp. 4603–4610, 2012.
21. C. Li, G. Hong, and L. Qi, "Nanosphere lithography at the gas/liquid interface: a general approach toward free-standing high-quality nanonets," Chemistry of Materials, vol. 22, no. 2, pp. 476–481, 2010.
22. S. B. Kim, W. W. Lee, J. Yi, W. I. Park, J. S. Kim, and W. T. Nichols, "Simple, large-scale patterning of hydrophobic ZnO nanorod

arrays," ACS Applied Materials and Interfaces, vol. 4, no. 8, pp. 3910–3915, 2012.

23. J. J. Dong, X. W. Zhang, Z. G. Yin et al., "Controllable growth of highly ordered ZnO nanorod arrays via inverted self-assembled monolayer template," ACS Applied Materials and Interfaces, vol. 3, no. 11, pp. 4388–4395, 2011.

24. A. Winkleman, B. D. Gates, L. S. McCarty, and G. M. Whitesides, "Directed self-assembly of spherical particles on patterned electrodes by an applied electric field," Advanced Materials, vol. 17, no. 12, pp. 1507–1511, 2005.

25. J. Aizenberg, P. V. Braun, and P. Wiltzius, "Patterned colloidal deposition controlled by electrostatic and capillary forces," Physical Review Letters, vol. 84, no. 13, article 2997, 2000.

26. N. D. Denkov, O. D. Velev, P. A. Kralchevsky, I. B. Ivanov, H. Yoshimura, and K. Nagayama, "Mechanism of formation of two-dimensional crystals from latex particles on substrates," Langmuir, vol. 8, no. 12, pp. 3183–3190, 1992.

27. A. S. Dimitrov and K. Nagayama, "Continuous convective assembling of fine particles into two-dimensional arrays on solid surfaces," Langmuir, vol. 12, no. 5, pp. 1303–1311, 1996.

28. D. Wang and H. Möhwald, "Rapid fabrication of binary colloidal crystals by stepwise spin-coating,"Advanced Materials, vol. 16, no. 3, pp. 244–247, 2004.

29. L. Wang, Y. Jiang, J. Luo et al., "Highly efficient and color-stable deep-blue organic light-emitting diodes based on a solution-processible dendrimer," Advanced Materials, vol. 21, no. 47, pp. 4854–4858, 2009.

30. T. R. Hebner, C. C. Wu, D. Marcy, M. H. Lu, and J. C. Sturm, "Ink-jet printing of doped polymers for organic light emitting devices," Applied Physics Letters, vol. 72, no. 5, pp. 519–521, 1998.

31. J. P. J. Markham, S.-C. Lo, S. W. Magennis, P. L. Burn, and I. D. W. Samuel, "High-efficiency green phosphorescence from spin-coated single-layer dendrimer light-emitting diodes," Applied Physics Letters, vol. 80, no. 15, pp. 2645–2647, 2002.

32. S. Tokito, T. Iijima, Y. Suzuri, H. Kita, T. Tsuzuki, and F. Sato, "Confinement of triplet energy on phosphorescent molecules

for highly-efficient organic blue-light-emitting devices," Applied Physics Letters, vol. 83, no. 3, pp. 569–571, 2003.

Chapter 2

Synthesis of ZnO Nanostructures by Microwave Irradiation Using Albumen as a Template

T. Prakash[1], R. Jayaprakash[1], G. Neri[2], and Sanjay Kumar[3]

[1]Nanotechnology Laboratory, Department of Physics, Sri Ramakrishna Mission Vidyalaya College of Arts and Science, Coimbatore, Tamil Nadu 641020, India

[2]Department of Electronic Engineering, Chemistry and Industrial Engineering, University of Messina, 98166 Messina, Italy

[3]Department of Physics, B. R. Ambedkar Bihar University, Muzaffarpur, Bihar 842001, India

ABSTRACT

ZnO nanostructures have been successfully prepared by a microwave irradiation method. The role of albumen as a template in addressing the size and morphology of ZnO has been investigated by X-ray diffraction (XRD), Fourier transform infrared spectroscopy (FT-IR), thermogravimetric analysis (TG-DTA), scanning electron microscopy (SEM), and transmission electron microscopy (TEM) techniques. A heterogeneous mixture of $Zn(OH)_2$ and ZnO was obtained in absence of albumen. Pure ZnO nanostructures, consisting of rod- and whisker-like nanoparticles embedded in a sheet matrix, were obtained in the presence of albumen. Optical and photoluminescence (PL) properties of the synthesized samples were also compared. Results obtained indicate that the microwave-assisted method is a promising low temperature, cheap, and fast method for the production of ZnO nanostructures.

INTRODUCTION

Recently, metal oxides which exhibit morphologies at the nanoscale (<100 nm) have become increasingly important. One of them is ZnO, a wide band gap n-type semiconductor (3.37 eV) with large exciton binding energy of (60 meV) at room temperature, which has attracted much attention due to its unique properties. ZnO nanoparticles have found numerous applications, such as in gas sensors [1], transparent electrodes [2], pH sensors [3], biosensors [4], acoustic wave devices [5], and UV photodiodes [6].

Up to now, a number of investigations have focused on the synthesis of nano- and microarchitectures through different methods like hydrothermal [7], sol-gel [8], electrochemical deposition [9], and vapor-phase process [10]. Previous studies on the microwave radiation synthesis of ZnO with a variety of morphologies, such as nanowires [11], nanorods [12], nanoneedles [13], hollow structures, and self-assembled architectures [14], were widely reported.

In this work, we synthesized ZnO nanostructures via a microwave irradiation method in the presence of albumen as a biotemplate. Microwave irradiation as a heating method has found a number of applications in chemistry. The utilization of microwave irradiation in the preparation of nanoparticles has been reported in recent years [15]. Compared to the conventional methods, the microwave synthesis has the advantages of producing small particle size metal oxides with high purity owing to short reaction time [15–19]. Due to the "in situ" mode of energy conversion, the microwave heating process is fundamentally different from conventional heating processes. Heat will be generated internally within the material, instead of originating from external sources. By means of this method, many functional materials and compounds with novel structures and properties have been obtained [20]. In recent studies, it has been shown that nanoparticles, such as Co_3O_4 [21], $Limn_{2-x}Cr_xo_4$ [22], SnO_2 [23] $Ce_{1-x}Gd_xO_{2-1/2x}$ [24], MgO [25], ZnO [26], and TiO_2 [27], can be prepared by the microwave synthesis process. Though the application of microwave irradiation in synthetic chemistry started in 1980s, only in the recent years, its use has provided a new, efficient, and environmentally benign methodology for the synthesis of various metal oxide nanoparticles of diverse morphologies and sizes. These radiations have unique properties like homogenous volumetric heating, which causes heating directly inside the sample, high reaction rate, selectivity, and increased product yield. Moreover, it is an energy saving process [14, 28, 29].

The process here proposed to synthesize ZnO nanostructures by a microwave-assisted process is simple, cheap, and fast and is characterized by a low synthesis temperature (130°C). As compared to conventional heating, microwaves cause the uniform distribution of temperature between the surface and the bulk material thereby leading to the fast formation of ZnO nanoparticles [30, 31]. Another advantage is that no postannealing process is necessary to obtain ZnO crystalline nanometer-sized structures compared to other microwave irradiation methods reported in the literature [32, 33]. To investigate the role of template in the synthesis of ZnO nanostructure by this method, a comparative study has been carried characterizing the products obtained in the presence or not of albumen as a biotemplate. The ZnO nanomaterials obtained were thoroughly characterized through X-ray diffraction (XRD), Fourier transform infrared spectroscopy (FT-IR), thermogravimetric analysis (TG-DTA), scanning electron microscopy (SEM), and transmission electron microscopy (TEM) techniques. Furthermore, their optical properties were assessed by UV-Vis DRS spectra and photoluminescence (PL) measurements.

EXPERIMENTAL PROCEDURE

Materials

Zinc nitrate $Zn(NO_3)_2$ and ammonia solution (NH_4OH) were supplied from (Merk, 98%) Mumbai, India. All the chemicals were of analytical grade and used as received without further purification. The albumen, extracted from white part of egg, was used for the synthesis. Double distilled water was used through the experiments.

Synthesis

The synthesis of ZnO nanostructures by the microwave irradiation method was carried out as follows. First, a 0.1 M of zinc hydroxyl solution was prepared by dissolving zinc nitrate ($Zn(NO_3)_2$) in deionized water. Then pH of the solution was maintained at 8 by adding liquid ammonia solution dropwise. The resulting product was filtered and washed with double distilled water and ethanol until it became free from impurities. The precipitate was irradiated for 5 minutes in household microwave (radiation frequency 2.45 GHZ, Power up to 1 KW) with convection mode, giving a white product. Finally the sample was dried at 130°C (sample A) for 5 hours.

The same experimental procedure was followed for the synthesis of ZnO nanostructures carried out in the presence of albumen (sample B), by adding drop by drop 5 mL of freshly extracted albumen, mixed with 25 mL of deionized water and stirred, to 0.1 M zinc hydroxyl solution.

Characterization

The sample microstructure was analyzed by X-ray diffraction (XRD) using a Bruker AXS D8 Advance instrument and the monochromatic CuK_1 wavelength of 1.5406⊠. The average crystalline size of the crystallites was evaluated using Scherrer's formula, $d = K\lambda / \beta\cos\theta$, where d is the mean crystalline size, k is a grain shape dependent constant (0.9), is the wavelength of the incident beam, is a Bragg reflection angle, and is the full width at half maximum (FWHM) of the main diffraction peak. The sample morphology was observed by scanning electron microscopy (SEM), using a JEOL 5600LV microscope at an accelerating voltage of 10 kV. High resolution transmission electron microscopy (HRTEM) and selected-area electron diffraction (SAED) were recorded on a Tecnai G20-stwin using an accelerating voltage of 200 kV. The Fourier transform infrared spectra (FT-IR) of the samples were recorded by using a Nicolet 5DX FTIR spectrometer. Thermal analysis was carried out by using a thermogravimetric and differential scanning calorimeter apparatus (TG-DSC Netzsch—Model STA 409PC). The analyses were carried out with a heating rate of 10°C/min in static air up to 800°C. The ultraviolet (UV) spectrum of the ZnO samples was recorded on a Perkin Elmer UV-visible DRS spectrophotometer. The room-temperature PL spectrum was performed on a spectrofluorometerinstrument (JY Fluorolog-FL3-11).

RESULTS AND DISCUSSION

Morphological and Microstructural Analysis

Thermogravimetric Analysis

The thermal behavior of ZnO samples synthesized has been first investigated by TG-DTA. Figure 1(a) shows the TG-DTA curves of sample A. The weight loss monitored in the range 25–600°C is low and has been calculated to be less than 3.3% in total. This weight loss is ascribed to the removal of absorbed water molecules and ethanol and the loss of hydroxyl ions. The final step at higher temperature is attributed to the burnout of organic species still remained in the dried powder. By focusing the attention on sample B, TGA indicated that the weight loss is higher (about 20% in total). The first weight loss in the temperature range between 25°C and 150°C is ascribed to the removal of adsorbed water molecules. The second weight loss in the range between 230°C and 520°C comes from the decomposition of the organic template, which is also confirmed by the exothermic peak at about 384°C in the DTA curve.

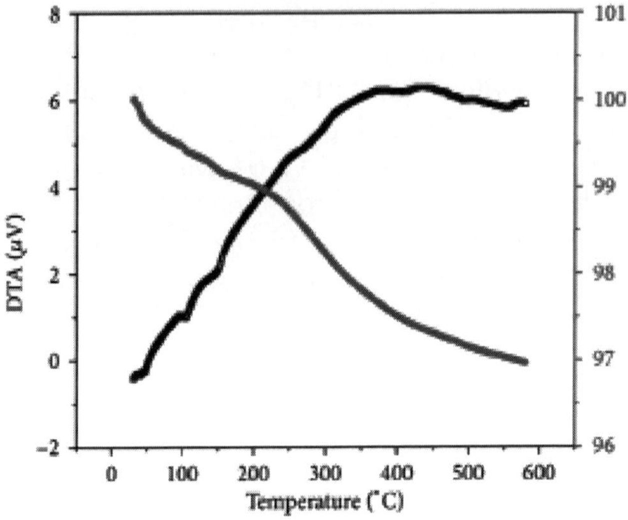

(a)

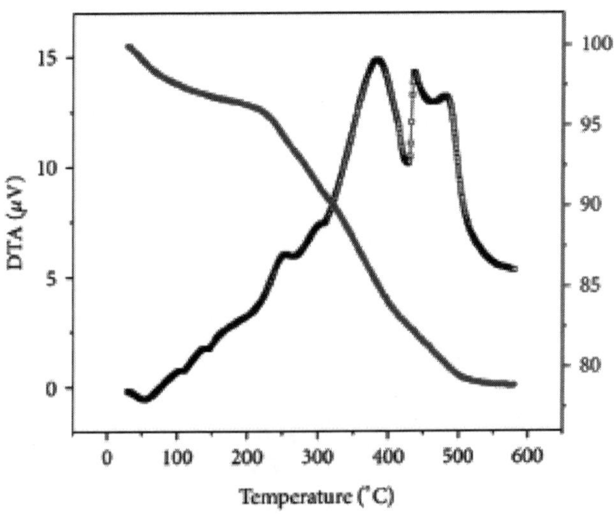

(b)

Figure 1: TG-DTA analysis of microwave synthesized samples: (a) sample A, (b) sample B.

X-Ray Diffraction

The phase identification of the synthesized ZnO nanostructures was determined by X-ray diffraction (XRD). Figures 2(a) and 2(b) show the XRD patterns of both samples. In Figure 2(a), XRD of sample A indicated that two phases, like ZnO hexagonal wurtzite phase structure (JCPDS number 897102) and $Zn(OH)_2$ orthorhombic phase structure (JCPDS number 890138), are present. On the basis of diffraction peaks intensity, ZnO is the predominant phase. According to Scherrer's equation, the average particle size of ZnO nanoparticles is about 29 nm.

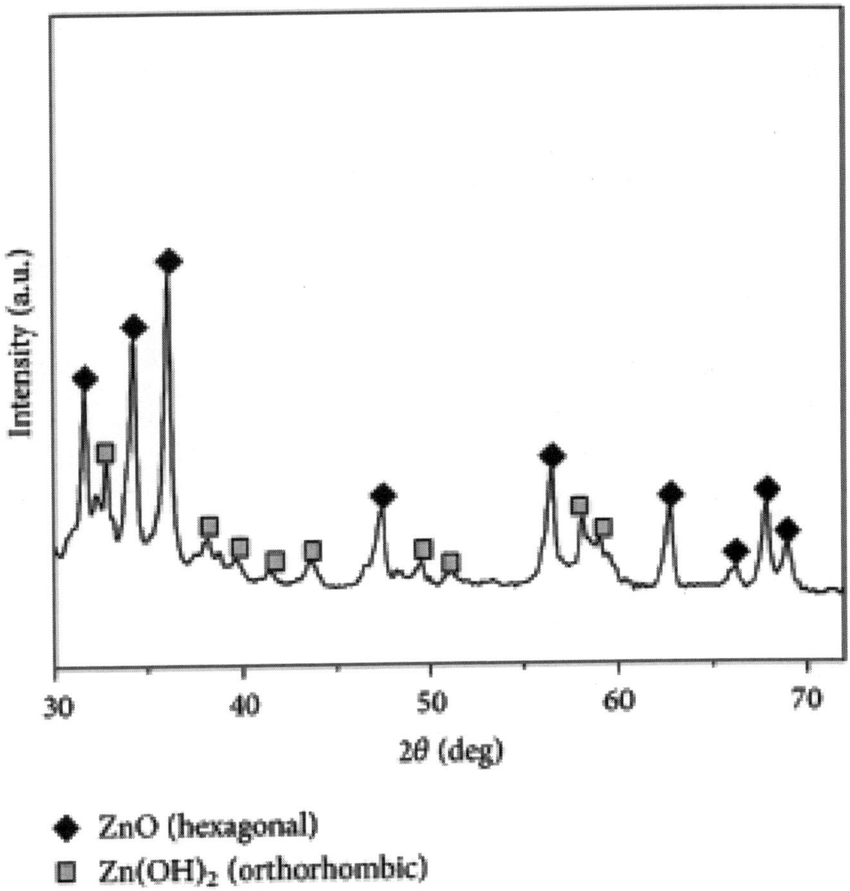

(a)

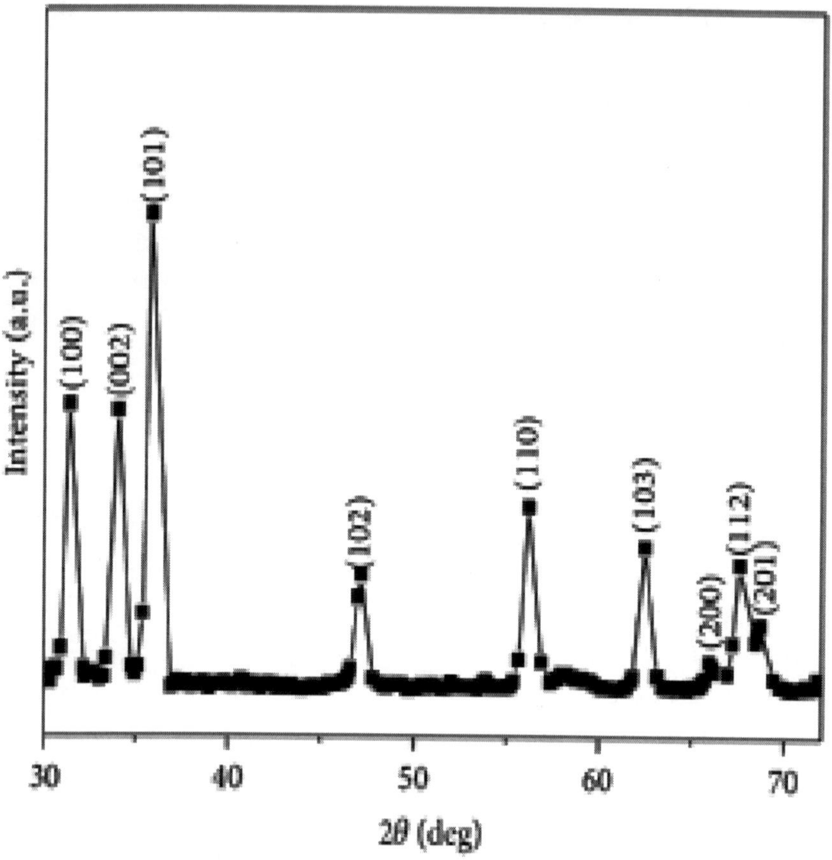

(b)

Figure 2: XRD analysis of microwave synthesized samples: (a) sample A, (b) sample B.

Figure 2(b) shows the XRD pattern of sample B. The diffraction planes (100), (002), (101), (102), (110), (103), (200), (112), and (201) in the patterns can be perfectly indexed to the hexagonal wurtzite phase structure (JCPDS number 897102). Also no diffraction peaks corresponding to Zn, $Zn(OH)_2$, and other impurities are observed. The value of lattice parameters (a= 3.2896 Å and c = 5.2624 Å), as calculated by the XRD data, is in good agreement with those reported by other authors [33]. The average particle size of ZnO is estimated to be around 14 nm.

XRD analysis indicates that the pure ZnO phase is obtained only in the presence of albumen. Furthermore, the average particle size of ZnO phase is markedly lower. It can be supposed that, in the formation processes in absence of albumen, the zinc hydroxide species was first produced step by step by the reaction between the Zn species and the hydroxyl ions coming from dissociation of hydrated ammonia molecules, and then in a consecutive reaction ZnO nanoparticles are formed. In the presence of template molecules, it appears that this second reaction is strongly promoted; so all initial zinc hydroxide species are transformed into ZnO.

Fourier Transform Infrared Spectroscopy

Figure 3(a) shows the typical FT-IR spectrum of sample A. The broad peak at 3706 cm^{-1} corresponds to the vibrational mode of O–H bond, indicating the presence of zinc hydroxide. The stretching mode of vibrations in asymmetric and symmetric C=O bonds that is observed at 1381 cm^{-1} is associated with residual organic impurity in the dried powder as suggested by TG-DTA analysis.

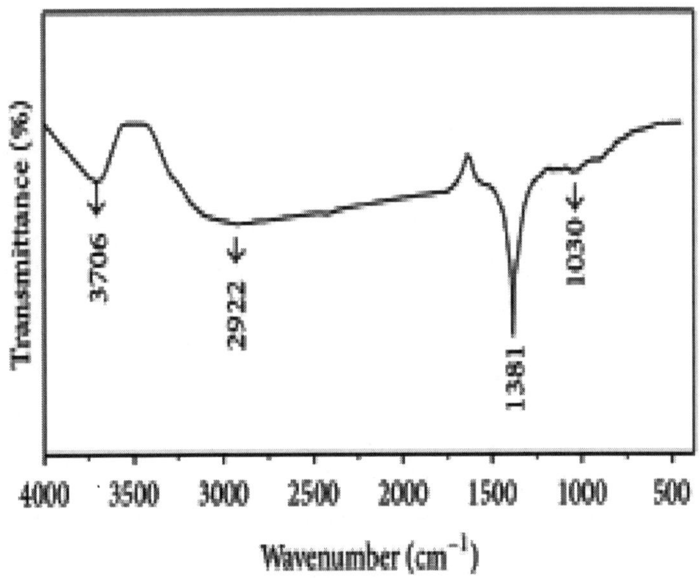

(a)

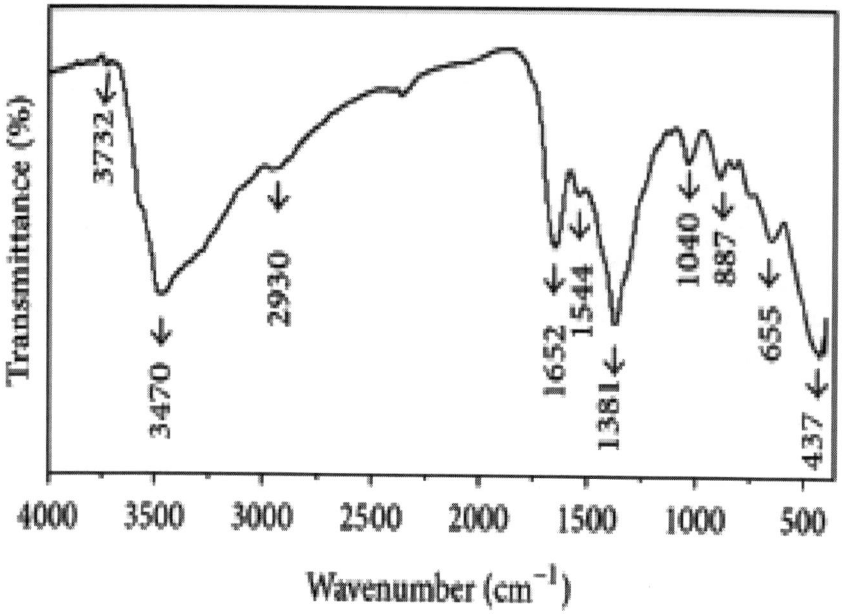

(b)

Figure 3: FT-IR spectra of microwave synthesized samples: (a) sample A, (b) sample B.

In the spectrum of sample B (Figure 3(b)), the appearance of a sharp band at 437 cm^{-1} in the FT-IR spectrum, characteristic for the Zn–O stretching vibration [34], confirms the presence of well-crystallized ZnO. Additionally broad absorption peaks centered at around 3470 cm^{-1} and 1652 cm^{-1} are caused by the O–H stretching of the absorbed water reabsorption during the storage of the sample in ambient air [35]. The absence of features at 3706 cm^{-1} is an indication of the absence of $Zn(OH)_2$ phase on this sample. In this sample are also seen the signatures of the organic template. The broad peak at 3470 cm^{-1} is in fact due to superimposition of N–H stretching of amide group at 3449 cm^{-1} of the biotemplate and O–H stretching mode of water molecules. Also some of the absorption bands in the range 1000–1600 cm^{-1} are due to C=O and C=C stretching vibrational modes of albumen, as well the weak peaks located at 2930 cm^{-1} are due to symmetric and asymmetric C–H bonds, respectively [36]. A very small band originated at 887 cm^{-1} is probably due to the carbonate moieties which are generally observed when FT-IR samples are measured in air [37, 38].

Scanning and Transmission Electron Microscopy

SEM and TEM images of sample A are reported in Figure 4. SEM analysis shows extensive sheet-like nanostructure where spherical-shaped particles are also present. TEM analysis evidences clearly this composite structure and the large diameter distribution of the spherical-shaped nanoparticles. No attempt has been made to assign the two phases present, ZnO and $Zn(OH)_2$, to each of these nanostructures. The thickness of the sheets and spherical particles is in the range of about 13–50 nm. The SAED pattern taken from one of the particles is shown in the inset, and it further confirms the crystalline nature of the sample. In addition, the EDS spectrum of the sample confirms the presence of Zn and O in sample without any other element contamination. The atomic percent ratio of O and Zn in dried sample is 41.15:58.85, which might indicate that there is a small content of oxygen vacancy defects on the surface of as-obtained product.

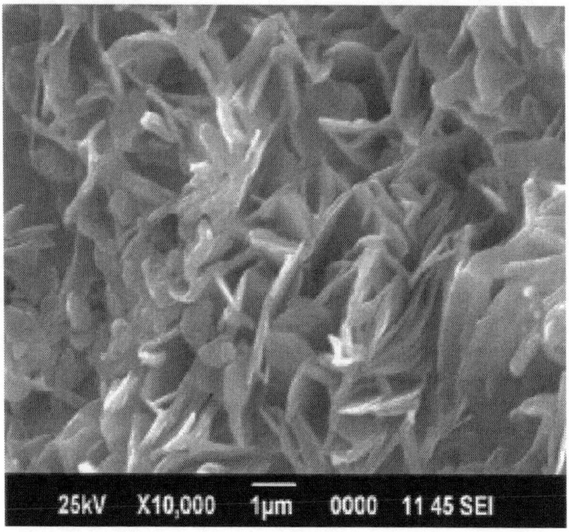

(a)

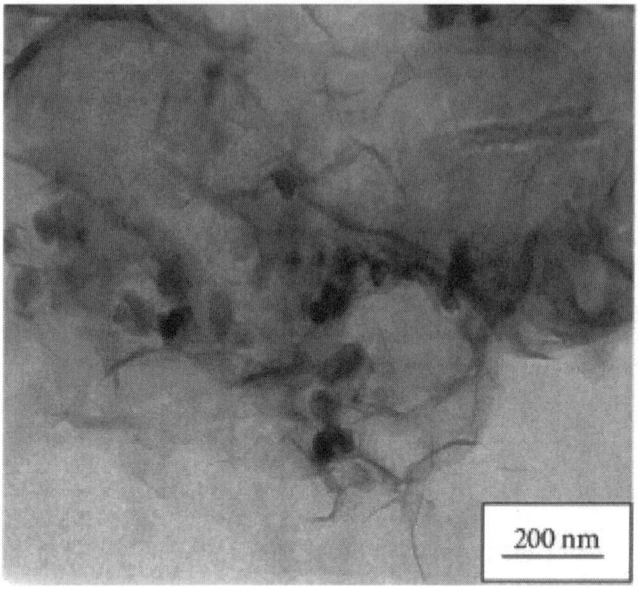

(b)

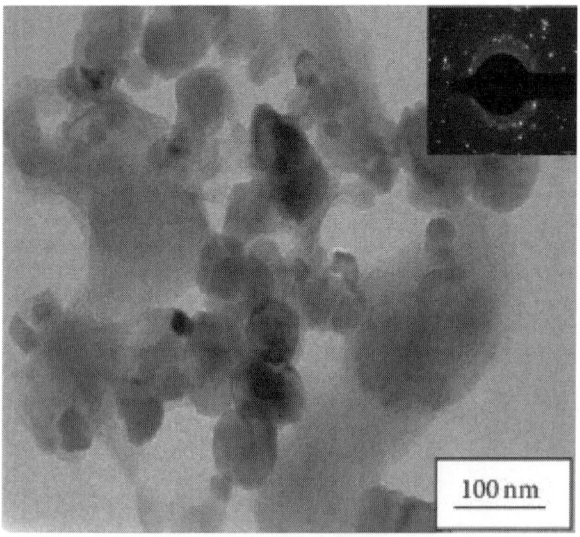

(c)

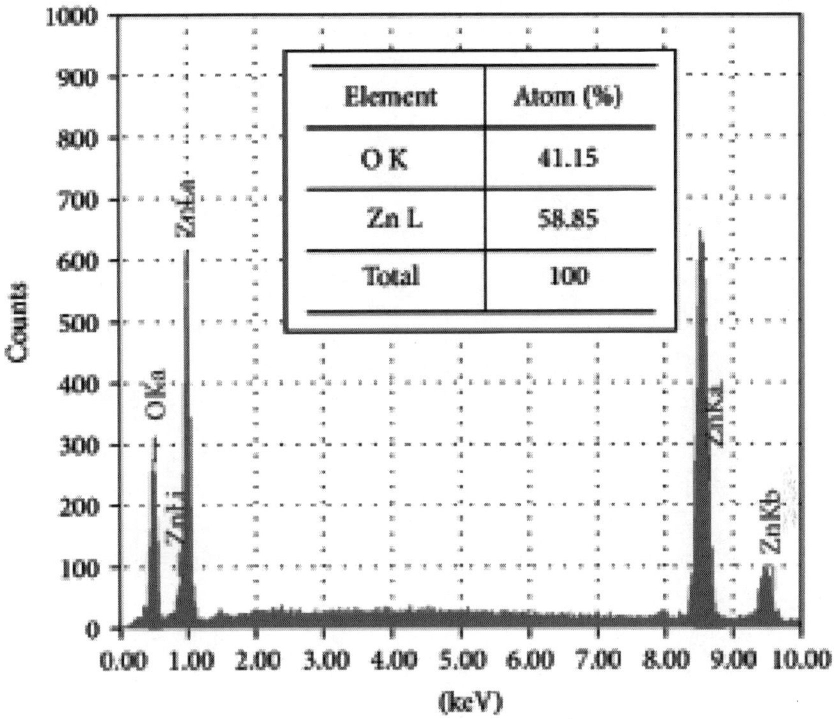

(d)

Figure 4: SEM and TEM micrographs of sample A. (a) SEM image, (b) and (c) TEM images of the sheet-like and spherical-like ZnO nanostructure, and (d) EDX analysis. Inset in Figure (c) shows the SAED pattern.

Figure 5 shows SEM and TEM images of sample B. SEM analysis shows that this sample is characterized by a more uniform surface sheet-like nanostructure. TEM analysis revealed the presence of nanowhiskers and nanorods of different length and diameter embedded within the sheet matrix. The thickness of the nanorods and nanowhiskers is in the range of about 10–57 nm. The SAED pattern of ZnO nanostructures further confirms the crystalline nature of the sample. EDS analysis reports that the atomic percent ratio of O and Zn in the dried sample is 85.59 : 14.11, which might indicate that there is a larger content of oxygen on the surface with respect to sample A.

(a)

(b)

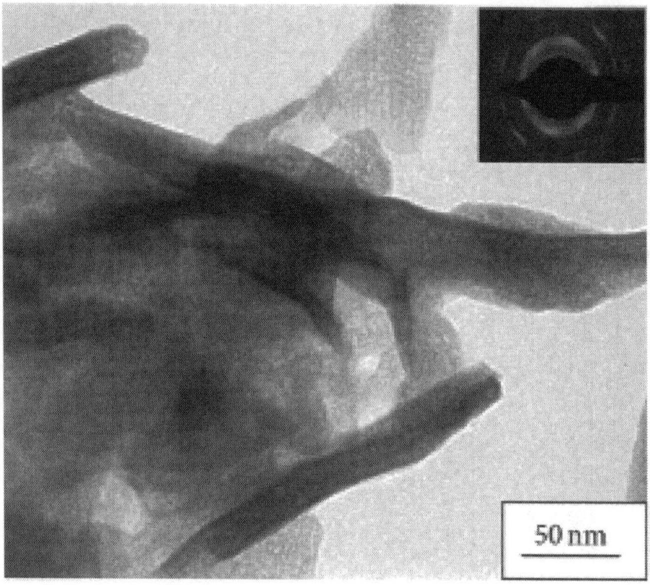

(c)

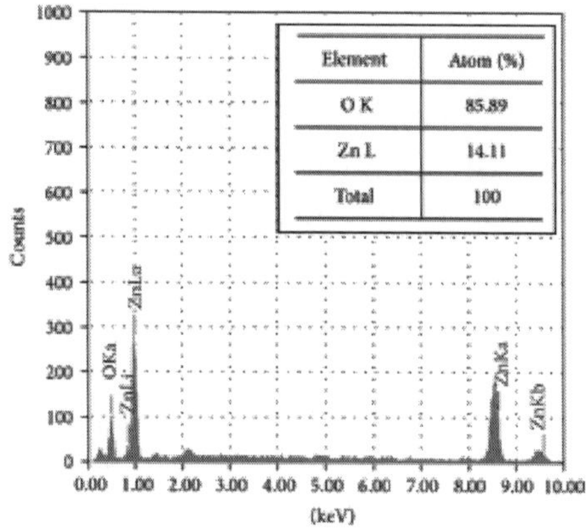

(d)

Figure 5: SEM and TEM micrographs of sample B. (a) SEM image, (b) and (c) TEM images of the whisker-like and rod-like ZnO nanostructure, and (d) EDX analysis. Inset in Figure (c) shows the SAED pattern.

Optical Properties

In order to determine the precise value of optical band gap UV-visible DRS of the ZnO nanoparticles in samples A and B, we used the optical absorption method on reflectance spectrum. The reflectance values were converted to absorbance by application of the Kubelka-Munk function [39–42]. The Kubelka-Munk theory is generally used for analyzing the diffuse reflectance spectra obtained from weakly absorbing samples. The Kubelka-Munk formula can be expressed by the following relation [43]:

$$K = \frac{(1-R)^2}{2R}, \tag{1}$$

where K is the reflectance transformed according to Kubelka-Munk and R is the reflectance (%).

The relationship $(k * h9)^{1/2} = f(h * 9)$ is shown in Figures 6(a) and 6(b). The E_9 value can be obtained by extrapolating the linear portion to the photon energy axis. The reflectance shoulder onset peaks are located at 393 nm and 386 nm corresponding to the band gaps of 3.15 eV and 3.21 eV, respectively. The E_9 values are smaller than that of 3.3 eV reported for single crystalline ZnO samples. On comparison with ZnO nanoparticles prepared by other methods, ZnO nanoparticles are obtained by this microwave irradiation method showing larger band gap energy. It is well known that the difference in the band gap is related to the presence of vacancies and dopants [44]. The red shift of the band gap energy also relates to structural morphologies, particle size, and surface microstructures [45]. Then, the higher value of band gap of the sample B synthesized in the presence of albumen can be attributed to a minor amount of defects, due essentially to its more ordered crystalline structure.

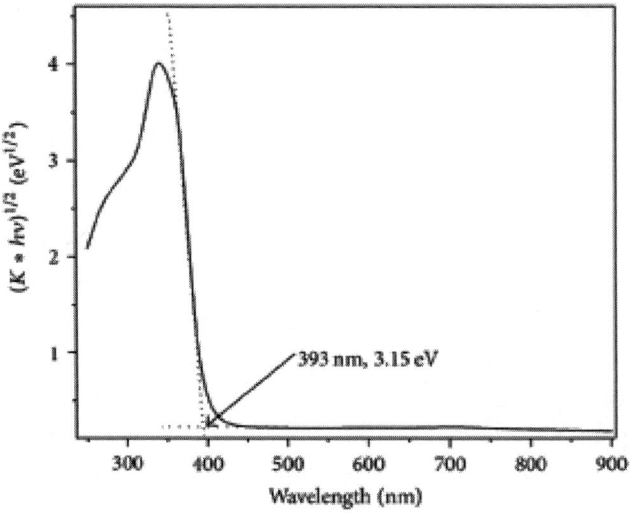

(a)

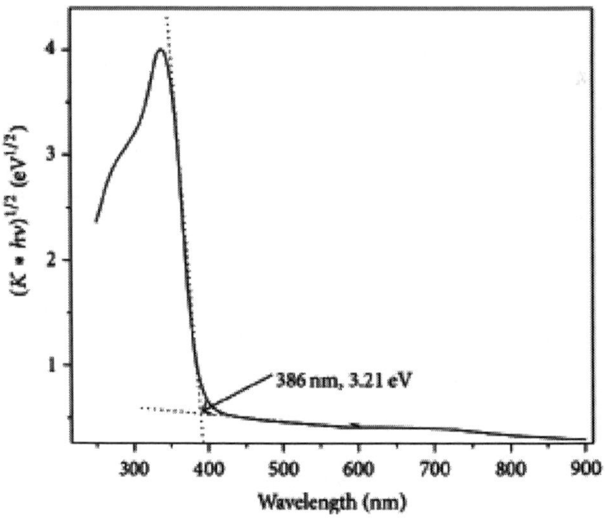

(b)

Figure 6: UV-Vis DRS of the typical ZnO nanoparticles: (a) sample A, (b) sample B.

Extrapolation of the linear portion to the photon energy axis to obtain the E_9 value is shown.

The study of the photoluminescence properties is interesting because it can provide valuable information on the quality, purity, and the structural properties of the material. Moreover, recording PL spectrum is of paramount importance for evaluating their optical characteristics for various applications. The PL spectra of samples A and B are depicted in Figures 7(a) and 7(b). On sample A, three emission peaks, a blue light at 400 nm, a blue-green around 464 nm, and a green light at around 520 nm, were recorded. The main strong UV emission at 400 nm corresponds to the near-band-edge emission, while the blue-green and green emission peaks at 464 nm and 520 nm are instead possibly associated with oxygen vacancies [46, 47]. Specifically, the blue-green transition at 464 nm is caused due to transition from the level of ionized oxygen vacancy to the valence band, while the green emission, called deep-level emission, occurs by the recombination of the photogenerated holes with singly ionized oxygen vacancies in ZnO. On sample B, an additional UV emission peak at 385 nm, due to the recombination of photogenerated electrons and holes, appears as a shoulder of the main peak at 400 nm. The higher intensity of near-band-edge emission peaks compared to green emission peaks is an indication of the good quality of ZnO nanostructures produced.

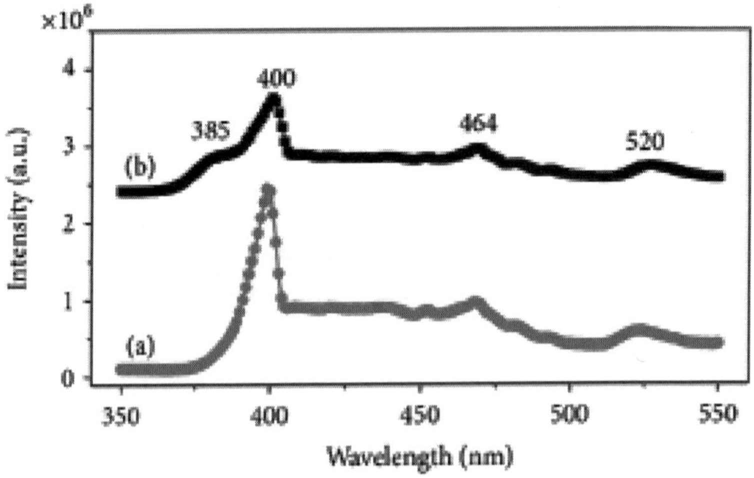

Figure 7: PL spectra of (curve a) sample A and (curve b) sample B.

CONCLUSIONS

The morphology, structure, and optical properties of ZnO nanostructures synthesized by microwave irradiation have been investigated in detail. The composition and morphology of the nanostructures are affected by the presence of albumen as a template. The optical band gap of the polycrystalline ZnO nanoparticles increases from 3.22 to 3.25 eV for the sample synthesized in the presence of albumen because of the defects' decrease. The good quality of ZnO nanostructures produced was also demonstrated by PL studies which evidenced a higher intensity of near-band-edge emission peaks compared to green emission peaks. In summary, results obtained indicate that the microwave-assisted method is a promising low temperature, cheap, and fast method for the production of ZnO nanostructures.

REFERENCES

1. B. B. Rao, "Zinc oxide ceramic semi-conductor gas sensor for ethanol vapour," Materials Chemistry and Physics, vol. 64, no. 1, pp. 62–65, 2000.
2. W. H. G. Horsthuis, "ZnO processing for integrated optic sensors," Thin Solid Films, vol. 137, no. 2, pp. 185–192, 1986.
3. Asif, Nur, Willander, and Danielsson, "Selective calcium ion detection with functionalized ZnO nanorods-extended gate MOSFET," Biosensors and Bioelectronics, vol. 24, no. 11, pp. 3379–3382, 2009.
4. C. Xia, N. Wang, L. Lidong, and G. Lin, "Synthesis and characterization of waxberry-like microstructures ZnO for biosensors," Sensors and Actuators B, vol. 129, no. 1, pp. 268–273, 2008.
5. Krishnamoorthy and Iliadis, "Development of high frequency ZnO/SiO2/Si Love mode surface acoustic wave devices," Solid-State Electronics, vol. 50, no. 6, pp. 1113–1118, 2006.
6. Liu, Zhang, Lu et al., "Fabrication and characterization of ZnO film based UV photodetector," Journal of Materials Science, vol. 20, no. 3, pp. 197–201, 2009.

7. Y. Wang and M. Li, "Hydrothermal synthesis of single-crystalline hexagonal prism ZnO nanorods,"Materials Letters, vol. 60, no. 2, pp. 266–269, 2006.
8. Lee, Easteal, Pal, and Bhattacharyya, "Evolution of ZnO nanostructures in sol-gel synthesis," Current Applied Physics, vol. 9, no. 4, pp. 792–796, 2009.
9. Wang, Liu, Chang, and Tang, "Synthesis of sulfur-doped ZnO nanowires by electrochemical deposition," Materials Science in Semiconductor Processing, vol. 10, no. 6, pp. 241–245, 2007.
10. J. Park, Y. Choi, and J. Park, "Synthesis of ZnO nanowires and nanosheets by an O2-assisted carbothermal reduction process," Journal of Crystal Growth, vol. 280, no. 1-2, pp. 161–167, 2005.
11. H. Cheng, J. Cheng, Y. Zhang, and Q. Wang, "Large-scale fabrication of ZnO micro-and nano-structures by microwave thermal evaporation deposition," Journal of Crystal Growth, vol. 299, no. 1, pp. 34–40, 2007.
12. J. Liu, J. Cao, Z. Li, G. Ji, and M. Zheng, "A simple microwave-assisted decomposing route for synthesis of ZnO nanorods in the presence of PEG400," Materials Letters, vol. 61, no. 22, pp. 4409–4411, 2007.
13. S. Cho, D. Shim, S. Jung, E. Oh, B. R. Lee, and K. Lee, "Fabrication of ZnO nanoneedle arrays by direct microwave irradiation," Materials Letters, vol. 63, no. 9-10, pp. 739–741, 2009.
14. Krishnakumar, Jayaprakash, N. Pinna, Singh, Mehta, and Phani, "Microwave-assisted synthesis and characterization of flower shaped zinc oxide nanostructures," Materials Letters, vol. 63, no. 2, pp. 242–245, 2009.
15. I. Y. Y. Bu, "Rapid synthesis of ZnO nanostructures through microwave heating process," Ceramics International, vol. 39, pp. 1189–1194, 2013.
16. S. T. Aruna and A. S. Mukasyan, "Combustion synthesis and nanomaterials," Current Opinion in Solid State and Materials Science, vol. 12, no. 3-4, pp. 44–50, 2008.
17. Mohebbi, Ebadzadeh, and Hesari, "Synthesis of nano-crystalline (Ni/NiO)-YSZ by microwave-assisted combustion synthesis method: the influence of pH of precursor solution," Journal of Power Sources, vol. 178, no. 1, pp. 64–68, 2008.

18. M. A. Bhosale, K. D. Bhatte, and B. M. Bhanage, "A rapid, one pot microwave assisted synthesis of nanosize cuprous oxide," Powder Technology, vol. 235, pp. 516–519, 2013.
19. P. Dinka and A. S. Mukasyan, "In situ preparation of oxide-based supported catalysts by solution combustion synthesis," Journal of Physical Chemistry B, vol. 109, no. 46, pp. 21627–21633, 2005.
20. Kasapoǧlu, Baykal, Köseoǧlu, and Toprak, "Microwave-assisted combustion synthesis of CoFe2O4 with urea, and its magnetic characterization," Scripta Materialia, vol. 57, no. 5, pp. 441–444, 2007.
21. L. Ai and Jiang, "Rapid synthesis of nanocrystalline Co3O4 by a microwave-assisted combustion method," Powder Technology, vol. 195, no. 1, pp. 11–14, 2009.
22. Y. Fu, Y. Su, and C. Lin, "Comparison of microwave-induced combustion and solid-state reaction for synthesis of LiMn2-xCrxO4 powders and their electrochemical properties," Solid State Ionics, vol. 166, no. 1-2, pp. 137–146, 2004.
23. A. Srivastava, Lakshmikumar, Srivastava, Rashmi, and K. Jain, "Gas sensing properties of nanocrystalline SnO2 prepared in solvent media using a microwave assisted technique," Sensors and Actuators B, vol. 126, no. 2, pp. 583–587, 2007.
24. Y. Fu, Y. Chang, and S. Wen, "Microwave-induced combustion synthesis and electrical conductivity of Ce1-xGdxO2-1/2x ceramics," Materials Research Bulletin, vol. 41, no. 12, pp. 2260–2267, 2006.
25. E. Esmaeili, A. Khodadadi, and Y. Mortazavi, "Microwave-induced combustion process variables for MgO nanoparticle synthesis using polyethylene glycol and sorbitol," Journal of the European Ceramic Society, vol. 29, no. 6, pp. 1061–1068, 2009.
26. C. Hwang and T. Wu, "Combustion synthesis of nanocrystalline ZnO powders using zinc nitrate and glycine as reactants - Influence of reactant composition," Journal of Materials Science, vol. 39, no. 19, pp. 6111–6115, 2004.
27. M. Selvakumar and D. K. Bhat, "Microwave synthesized nanostructured TiO_2-activated carbon composite electrodes for supercapacitor Selvakumar," Applied Surface Science, vol. 263, pp. 236–241, 2012.

28. I. Bilecka, P. Elser, and M. Niederberger, "Kinetic and thermodynamic aspects in the microwave-assisted synthesis of ZnO nanoparticles in benzyl alcohol," ACS Nano, vol. 3, no. 2, pp. 467–477, 2009
29. P. Zhu, J. Zhang, Z. Wu, and Z. Zhang, "Microwave-assisted synthesis of various ZnO hierarchical nanostructures: effects of heating parameters of microwave oven," Crystal Growth and Design, vol. 8, no. 9, pp. 3148–3153, 2008
30. N. F. Hamedani, A. R. Mahjoub, A. A. Khodadadi, and Y. Mortazavi, "Microwave assisted fast synthesis of various ZnO morphologies for selective detection of CO, CH4 and ethanol," Sensors and Actuators B, vol. 156, no. 2, pp. 737–742, 2011.
31. J. ZHU, J. ZHANG, H. ZHOU, W. QIN, L. CHAI, and Y. HU, "Microwave-assisted synthesis and characterization of ZnO-nanorod arrays," Transactions of Nonferrous Metals Society of China (English Edition), vol. 19, no. 6, pp. 1578–1582, 2009.
32. Savary, Marinel, Colder, Harnois, Lefevre, and Retoux, "Microwave sintering of nano-sized ZnO synthesized by a liquid route," Powder Technology, vol. 208, no. 2, pp. 521–525, 2011.
33. L. C. Nehru, V. Swaminathan, and C. Sanjeeviraja, "Rapid synthesis of nanocrystalline ZnO by a microwave-assisted combustion method," Powder Technology, vol. 226, pp. 29–33, 2012.
34. Al-Hajry, A. Umar, Hahn, and Kim, "Growth, properties and dye-sensitized solar cells-applications of ZnO nanorods grown by low-temperature solution process," Superlattices and Microstructures, vol. 45, no. 6, pp. 529–534, 2009.
35. T. Prakash, R. Jayaprakash, D. Sathya Raj et al., "Sensing properties of ZnO nanoparticles synthesized by using albumen as a biotemplate for acetic acid monitoring in aqueous mixture," Sensors and Actuators B, vol. 176, pp. 560–568, 2013.
36. Vafaee and M. S. Ghamsari, "Preparation and characterization of ZnO nanoparticles by a novel sol-gel route," Materials Letters, vol. 61, no. 14-15, pp. 3265–3268, 2007.
37. A. Umar, Rahman, M. Vaseem, and Y. Hahn, "Ultra-sensitive cholesterol biosensor based on low-temperature grown ZnO nanoparticles," Electrochemistry Communications, vol. 11, no. 1, pp. 118–121, 2009.

38. A. Umar, Rahman, Al-Hajry, and Hahn, "Highly-sensitive cholesterol biosensor based on well-crystallized flower-shaped ZnO nanostructures," Talanta, vol. 78, no. 1, pp. 284–289, 2009.
39. A. E. Morales, E. S. Mora, and U. Pal, "Use of diffuse reflectance spectroscopy for optical characterization of un-supported nanostructures," Revista Mexicana De Fisica S, vol. 53, pp. 18–22, 2007.
40. Senthilkumar, Vickraman, and Ravikumar, "Synthesis of fluorine doped tin oxide nanoparticles by sol-gel technique and their characterization," Journal of Sol-Gel Science and Technology, vol. 53, no. 2, pp. 316–321, 2010.
41. F. Yakuphanoglu, "Electrical characterization and device characterization of ZnO microring shaped films by sol-gel method," Journal of Alloys and Compounds, vol. 507, no. 1, pp. 184–189, 2010.
42. Yakuphanoglu, Mehrotra, Gupta, and M. Oz, "Nanofiber organic semiconductors: the effects of nanosize on the electrical charge transport and optical properties of bulk polyanilines," Journal of Applied Polymer Science, vol. 114, no. 2, pp. 794–799, 2009.
43. J. Tauc, R. Grigorovici, and A. Vancu, "Optical properties and electronic structure of amorphous germanium," Physica Status Solidi B, vol. 15, pp. 627–637, 1966.
44. Wang, Chen, and Yang, "Microstructure and optical properties of polycrystalline ZnO films sputtered under different oxygen flow rates," Journal of Alloys and Compounds, vol. 488, no. 1, pp. 232–237, 2009.
45. A. K. Zak, M. E. Abrishami, W. H. Abd Majid, R. Yousefi, and S. M. Hosseini, "Effects of annealing temperature on some structural and optical properties of ZnO nanoparticles prepared by a modified sol-gel combustion method," Ceramic International, vol. 37, pp. 393–398, 2011.
46. P. Yang, H. Yan, S. Mao et al., "Controlled growth of ZnO nanowires and their optical properties,"Journal of Advanced Functional Materials, vol. 12, pp. 323–331, 2002.
47. P. Jiang, J. Zhou, H. Fang, C. Wang, Z. L. Wang, and S. Xie, "Hierarchical shelled ZnO structures made of bunched nanowire arrays," Advanced Functional Materials, vol. 17, no. 8, pp. 1303–1310, 2007.

Chapter 3

Prospects of Nanostructure-Based Solar Cells for Manufacturing Future Generations of Photovoltaic Modules

N. Gupta[1], G. F. Alapatt[1], R. Podila[2], R. Singh[1], and K. F. Poole[1]

[1]Center for Silicon Nanoelectronics and Holcombe, Department of Electrical and Computer Engineering, Clemson University, Clemson, SC 29634-0915, USA

[2]Department of Physics and Astronomy, Clemson University, Clemson, SC 29634-0915, USA

ABSTRACT

We present a comprehensive review on prospects for one-, two-, or three-dimensional nanostructure-based solar cells for manufacturing the future generation of photovoltaic (PV) modules. Reducing heat dissipation and utilizing the unabsorbed part of the solar spectrum are the key driving forces for the development of nanostructure-based solar cells. Unrealistic assumptions involved in theoretical work and the tendency of stretching observed experimental results are the primary reasons why quantum phenomena-based nanostructures solar cells are unlikely to play a significant role in the manufacturing of future generations of PV modules. Similar to the invention of phase shift masks (to beat the conventional diffraction limit of optical lithography) clever design concepts need to be invented to take advantage of quantum-based nanostructures. Silicon-based PV manufacturing will continue to provide sustained growth of the PV industry.

INTRODUCTION

Globally the photovoltaic (PV) market grew 110% last year and reached a record 5.95 GW in 2008 [1]. It is also predicted that by 2012 more than 20 GW PV systems will be installed. Gartner has predicted an increase in PV industry revenue to $34 billion within the next five years from $15.8 billion in 2008 [2]. Currently, wafer-based silicon (single crystal, poly crystalline, and multicrystalline) solar cells and thin film solar cells based on amorphous silicon, CdTe, $CuInGaSe_2$, and III–V semiconductors dominate PV manufacturing [3]. For future generations of solar cells a number of approaches are being explored [4]. An examination of the current literature shows that most of these solar cells are based on one-, two-, or three-dimensional structures with nanodimensions (<100 nm). The R&D work related to the nanostructures falls under the buzz word of "nanotechnology" and there are questionable claims on the scientific facts [5]. As an example, the publication of a research paper on multiexciton generation in colloidal silicon nanocrystals [6] motivated the author of [7] to claim that silicon nanocrystals-based solar cells can generate two electrons from one photon, and single junction solar cell efficiencies as high as 40% are possible. The fact is that multiple exciton generation phenomena observed in nanocrystals or quantum dots have no direct relevance to the operation of the solar cell. The incorrect assumption that multiple electron hole pairs (EHPs) generation will improve the solar cell efficiency ignores the fact that the collection of the carriers is a necessary part of the solar cell operation. This can impact the further direction of basic research that is generally funded by public money [8] as well as investment of private money for establishing new photovoltaic businesses. In the broader sense of green electricity generation our objective is to examine critically all the current PV approaches that are under investigation.

PV manufacturing has its origin in semiconductor manufacturing and most of the thin film PV manufacturing (use of large glass substrate) has its origin in liquid crystal display (LCD) manufacturing. Based on an examination of the fundamentals of materials, devices, and the experience of semiconductor manufacturing, LCD manufacturing, and related manufacturing, it is possible to examine the prospects of incorporating nanostructure-based solar cells into PV manufacturing for next generation solar cells. Recently, a number of review papers on applications of nanostructures in solar cells have been published [3, 9–18]. However, none of these papers address the issues related to manufacturing of next generations of PV modules. The objective of this paper is to critically examine the published work and point out the suitability or nonsuitability of the particular nanostructure for future generations of photovoltaic module manufacturing. In the following section, we provide the background material which is essentially required to make PV systems a significant source of electricity generation. In the third section, we examine the theoretical and experimental results associated with the materials and devices of nanostructure-based future generation of solar cells. The discussion of the paper is presented in Section 4. Finally, the paper is concluded in Section 5.

BACKGROUND MATERIAL

In order to create huge PV markets, the following selection criterion must be used in the selection of appropriate technology: (i) no material supply constraint, (ii) low cost of ownership, (iii) low production cost, (iv) prospects of further cost reduction, and (v) green manufacturing with no environmental safety and health issues. In a previous publication [3], we have discussed these points in depth however for completeness key points are briefly discussed in this section.

No Material Supply Constraints

Just after oil embargo of 1973 [19], a number of semiconductor materials were proposed as the potential materials for manufacturing solar cells. In 1980, one of us published an article on the economic requirements for new materials for solar cells and predicted that silicon is the best material for large-scale PV manufacturing [20]. Over the last 29 years our prediction has been correct, since more than 90% solar cells manufactured in 2008 are made out of silicon. Thus it is quite relevant to go back to our 1980 paper and rewrite the rules for selecting photovoltaic materials. The key guidelines from 1980 publications [20] are given in the following paragraphs.

When one considers the availability of materials at reasonable cost, one has to establish the rate of production required for the application and compare it to the current world rate of production as well as to the possible total availability of the material in the earth's crust. As shown in Figure 1, a number of elements used in solar cell manufacturing have very low level of reserve [21]. In order that the use of the material in the photovoltaic application does not drive up the cost of the material, the additional uses for photovoltaic applications must be less than 10% of the world production (study of world production rates of minerals shows that historically the production rate of any mineral has rarely grown faster than 10% per year). For example, due to high demand of poly-silicon, price of poly-silicon went high in 2008 due to increased demand and poor supply. Fortunately, due to abundance of the raw material, the supply chain of poly-silicon has improved again. Recently, iSuppli has reported that due to an increase in production of poly-silicon by 112% in 2009, price will drop by half [22].

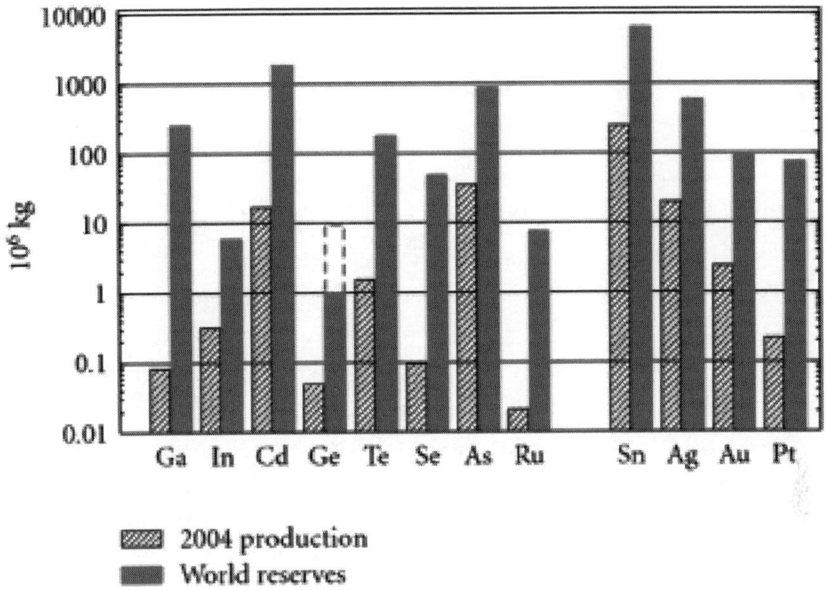

Figure 1: Estimated 2004 annual production levels and material reserves for various materials used in PV industry (from [21]).

Low Cost of Ownership (COO) for Manufacturing

Similar to the semiconductor manufacturing and LCD manufacturing industry, photovoltaic manufacturing also needs huge capital investments. Ideally for a given technology, the capital expenditures should be as low as possible. However, capital expenditure is only one component in the cost of ownership. Since the manufacturing of a PV module consists of a number of processing steps, the underutilization of equipment adds extra cost. Over the life time of the manufacturing facility, the equipment reliability, utilization, and yield of the PV modules have a greater impact on cost of ownership than initial purchase cost. The cost of ownership can be expressed by following expression [23]:

$$\text{COO} = \frac{\text{CF} + \text{CV} + \text{CY}}{\text{TPT} * \text{Y} * \text{U}}, \tag{1}$$

where COO: cost of ownership, CF: fixed cost, CV: variable cost, CY: cost due to yield loss, TPT: throughput, Y: composite Yield, U: utilization.

The use of a roll-to-roll (RTR) process (similar to the way in which newspapers are printed) is a driving force to develop PV manufacturing processes with low capital expenditures. The success or failure of RTR and other similar processes depends on the yield of each process step and resulting PV module yield. The above expression can be used to judge the suitability or nonsuitability of a low-cost capital expenditure manufacturing process versus alternate high-cost capital expenditure manufacturing.

In the selection of appropriate manufacturing process technology, process variability is one of the important issues. In other words the power out of each device in a module and the power output of each module in a system must be as identical as possible. Due to series and parallel connections (to get desired voltage and current) devices or modules with minimum voltage or current will dictate the power output of the system. A sophisticated and expensive advanced process control (APC) will provide lower cost of ownership than a low-cost statistical process control for a given process used in the PV module manufacturing. For one-dimensional nanostructure such as single-wall carbon nanotubes, we have demonstrated the need for a high level of control of the process variation which is far beyond that which is currently achievable through conventional methods [24]. Process variability is one of the most critical issues for nanostructure-based integrated circuit manufacturing. The same issues are also true for solar cells based on nanostructures. As shown in Figure 2, for nanostructures-based solar cells the process window is replaced by the process points. Relative to currently manufactured solar cells, the COO will be higher for nanostructure-based solar cells.

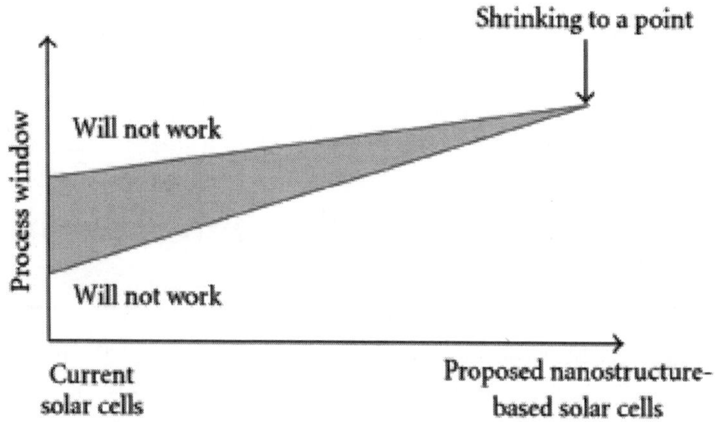

Figure 2: Shrinking process window for ultra-small structures.

Low Production Cost

For reducing the production cost of PV modules, it is necessary to use larger silicon wafers or larger substrates rather than the smaller ones. For product cost reduction, this technique has been very successful for IC and display industry. Other than the efficiency of PV modules, energy consumed in the manufacturing process, materials, gases and substrates cost, automation and throughput, yield is a very important factor in the overall cost. Of course factors such as low labor cost and low cost of electricity also affect the cost of PV modules. The typical manufacturing cost relationship should be used to size the appropriate production capacity of the manufacturing process. An absolute control on the supply chain and ultra-large-scale volume manufacturing are the principal requirements to bring PV technology to a level where generation of electricity is cheaper than other energy sources.

Prospects for Further Cost Reduction

Other than increasing the efficiency of PV module, the manufacturing process should be capable of further cost reduction. For cost reduction of the PV modules, the change of substrate size with time by the IC and display industry is the perfect choice for PV industry. Currently some amorphous silicon manufacturers are using glass substrates of the size of 5.7 m^2. Similar to LCD industry these manufacturers can increase the glass size to Gen 10 substrates with the area equal to 8.7 m^2. Without inventing any new process, the Gen 10 substrate size will provide cost reductions similar to LCD industry.

Environmental Safety and Health (ESH) Issues and Green Manufacturing

As compared to other method of electricity generation, PV technologies have distinct environmental advantages for generating electricity over conventional technologies. Electricity generation by PV systems does not produce any noise, toxic-gas emissions, or greenhouse gases. Photovoltaic energy not only can help meet the growing worldwide demand for electricity, but it can do so economically without any environmental concerns. Compared to burning coal, every gigawatt-hour of electricity generated by photovoltaics would prevent the emission of about 10 tons of sulfur dioxide, 4 tons of nitrogen oxides, 0.7 tons of particulates, and up to 1000 tons of carbon dioxide.

The techniques used in PV manufacturing are quite similar to $350 billion semiconductor industry. During the last half century the semiconductor industry has evolved to the point that in spite of the use of a number hazardous chemicals, tools and processes exist for the safe handling of such chemicals, materials, and gases. The PV industry has adopted handling of materials chemicals, and gases in a way similar to the semiconductor industry. Thus from environmental, occupational health, and safety point, there is no concern. The only material that poses a real concern is the use of Cadmium (Cd) in CdTe-based solar cells. According to the Occupational Safety and Health Administration (OSHA) of USA, Cadmium is considered both toxic and a lung carcinogen. OSHA considers all Cd compounds (including CdTe) to be toxic. For further reading of health problems related to the use of Cd in solar cells interested reader is referred to our earlier publication [3].

CRITICAL EXAMINATION OF PROPOSED NANOSTRUCTURES FOR SOLAR CELL APPLICATIONS

The nanostructures involve length scale of approximately 1–100 nm range. In this paper we will not differentiate between the approaches used in the fabrication of these structures. The issue of "bottom up" (atom-by-atom approach to build the required < 100 nm dimension) or "top down"(current lithography techniques used to reduce material dimension < 100 nm) approach will be dictated by throughput and cost related manufacturing issues. It has been known for long time that the properties of ultra-thin films differ from the bulk values. The invention of scanning tunneling microscope in 1981 and the advancements in the silicon integrated circuits as well as other related semiconductor products (e.g., improved detectors with high signal-to-noise ratio) have helped the development of instruments that have the ability to manipulate and analyze material on the atomic scale. Research in the nanodimension material in the last decade has established that the properties of nano-dimension material differ from the corresponding thin film materials. As shown in Figure 3, at nanoscales the properties of material depend on the quantum confinements. The one-to three-dimensional nanomaterials provide unique opportunities to optimize a number of physical, chemical, biological, mechanical, electrical, optical, and thermal properties of interest for future photovoltaic devices. The challenge is to integrate these unique properties of nanostructures and enhance the performance, reliability, and yield of photovoltaic devices for manufacturing next generation of photovoltaic modules.

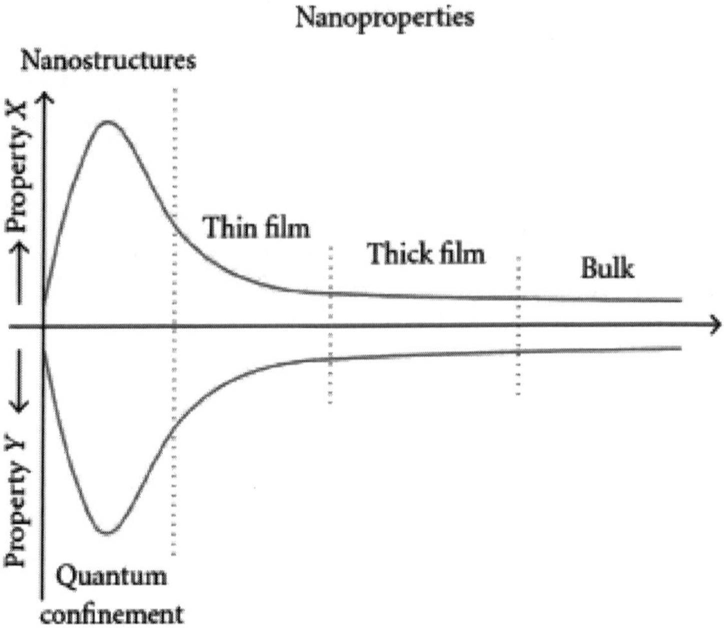

Figure 3: Change in properties of any material due to its physical size.

In 1961, Shockley and Queisser calculated thermodynamic limit of maximum efficiency for a p-n junction silicon (1.1 eV energy bandgap) solar cell to be around 33% under AM 1.5 solar illumination at 300^{0}K [25]. The energy associated with photons greater than the energy band gap of silicon is converted into heat. As shown in Table 1, heat loss is mainly responsible for lower efficiency of silicon solar cells. The unabsorbed energy loss can be partially recovered by using the concept of tandem solar cells where higher and lower energy band gap solar cells are connected optically in series. The concept of tandem solar cells is already used for manufacturing amorphous silicon solar cells and III–V-based high efficiency concentrator solar cells. Since a major part of the incident solar energy is lost as heat, maximum efforts have been made to prevent incident solar energy getting converted to heat. These losses are the driving force to build new approaches to utilize the energy losses. As mentioned earlier, structures of size 1 nm to 100 nm are proposed to make solar cells more efficient. The use of nanostructures has been suggested to build new highly efficient solar cells considering the following key points. (1) Hot carriers generated by absorbing photons of energy more than the energy band gap can be employed to generate additional current and or voltage, and (2) energy states in the energy band gap can be used to trap carriers generated from photon of energy lower than the energy band gap. Various approaches based on the one-, two-, and three-dimensional nanostructures (quantum dots and semiconductor nanocrystals) are currently being pursued to utilize the energy losses listed in Table 1. In the following subsections, we have examined the theoretical and experimental results of various nanostructures reported in the open literature.

Table 1: Various solar energy losses in single junction silicon (1.1 eV energy band gap) solar cell under AM 1.5 solar spectra at 300°K temperature [13]

Photo-electro energy conversion	**32.9**
Unabsorbed energy losses	18.7
Heat loss	46.8
Other losses	1.6

Proposed Use of Hot Carriers to Boost up Solar Cell Efficiency

There have been numerous publications discussing the potential use of hot carriers for solar cell efficiency enhancement [6, 9, 10, 12, 26–31]. Hot carrier solar cells claim to convert excess photon energies into stable form before the excited states equilibrate with the environment. In semiconductors, excess energy appears as the kinetic energy of the carriers where as in molecules it becomes a phonon and is dissipated as heat. As shown in Figures 4(a)-4(b), when the incoming photon has excess energy (hv)>E_g, electrons are excited and the thermalization occurs through carrier-phonon scattering. Excess energy is delivered to the carriers of lesser effective mass. Thus for the highest efficiency of the solar cell, excess photon energies must be delivered to the bands capable of sustaining the hot carriers. The following are the proposed ways for extraction of the excess energy.

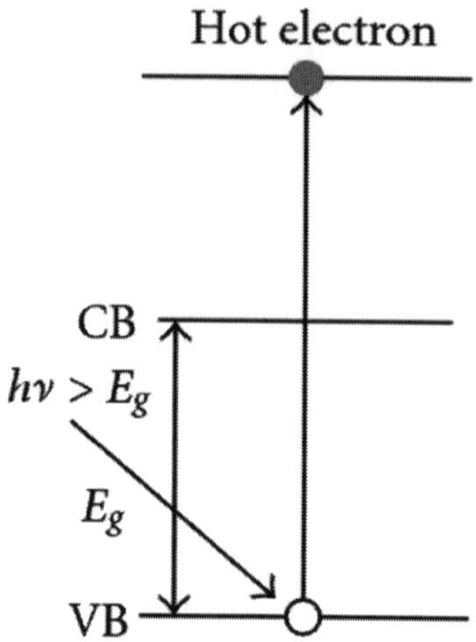

(a)

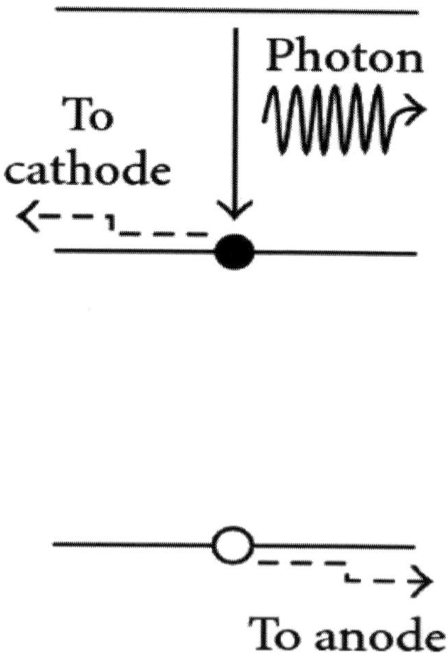

(b)

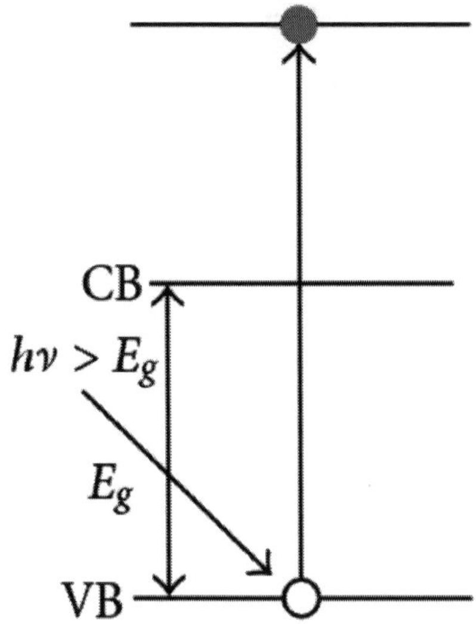

(c)

(d)

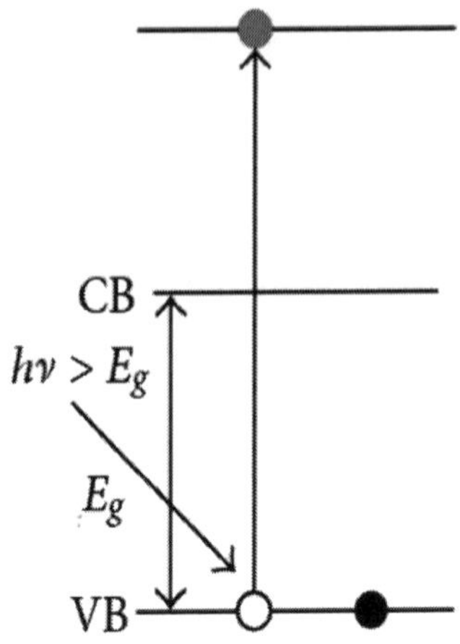

(e)

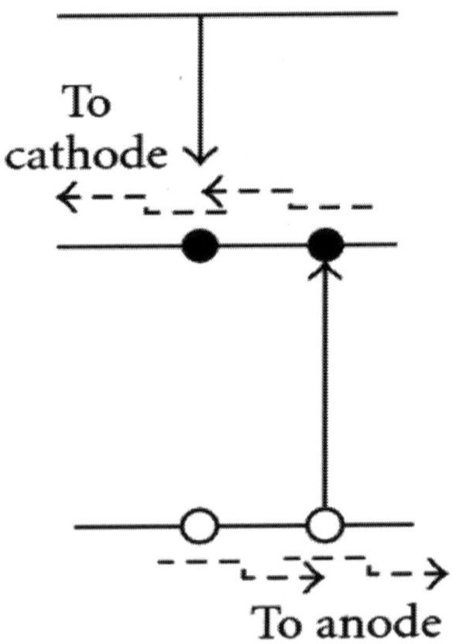

(f)

Figure 4: (a) A photon with energy (hv) greater than band gap energy (E_g) strikes on solar cell and excites an electron to Conduction Band

(CB) from Valence Band (VB); (b) the extra energy (hv–E_g) is released as phonon; (c) a photon with energy (hv) greater than band gap energy (E_g) strikes on solar cell and excites an electron to CB from VB; (d) the extra energy (hv–E_g) is used to produce higher voltage by hv–E_g ; (e) a photon with energy (hv) more than band gap energy (E_g) strikes on solar cell and excites an electron; (f) extra energy (hv–E_g) is used to excites one more electron from VB to CB.

- Remove the carrier from the device before it cools and obtain a higher energy electron and thus getting a higher output voltage (Figures 4(c)–4(d)). Theoretically for this to be feasible, the rate of cooling of hot electrons must be lower than the rate of hot carrier separation, transport, and collection at contacts [30].
- Make the excited "hot" electron spend its extra energy to excite another electron from valence band and thus getting a higher output current (Figures 4(e)–4(f)). This is known as impact ionization and can occur only if the rate of impact ionization is more than the relaxation processes which include generation of phonon and Auger recombination. [30].

The photocurrent generation in semiconductor p-n junction of solar cell is shown in Figure 5. When a photon with energy more than band gap of absorber material hits p-n junction, it creates an electron hole pair and due to the inbuilt electric field, photoproduced carriers are transported to the electrodes. Thus for making a solar cell, the basic requirements are an absorber material and a junction with electric field that assists transport of the photogenerated carriers to the electrodes. There is also a probability that photogenerated carriers can recombine before reaching the electrodes. Therefore for crystalline and polycrystalline solar cells, the charge separation time of photo generated carriers must be smaller than the minority carrier lifetime.

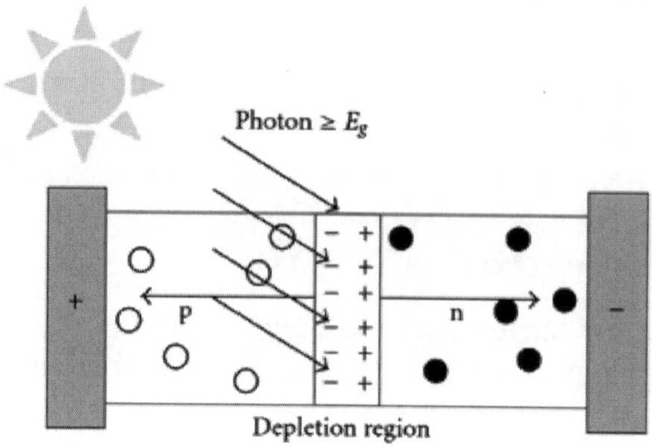

Figure 5: Basic p-n junction solar cell and charge transport phenomena.

For the bulk materials, rate of impact ionization or multiple carrier excitations (MCEs) is slower than the rate of thermal cooling of hot carriers through intraband relaxations processes [9]. Therefore various groups started to investigate the use of quantum dots for utilizing hot carrier effects [9, 29, 32]. However, to date no one has provided experimental data to prove that for quantum dots the rate of multiple carrier excitation is faster than the thermal cooling rate of hot carriers. Also, the Auger recombination rate and cooling rate depend on the surface area of the nanostructure as shown in Figure 6 [13]; for example, the Auger recombination is more in nanorods than nanodots [9]. Results shown in Figure 7 indicate that not only carrier relaxation time decreases, photo generated exciton life time also decreases with the size of nanocrystal [13].

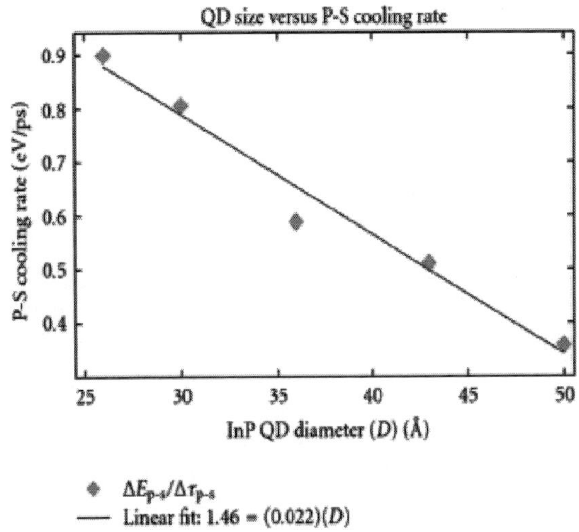

Figure 6: Carrier cooling rate increases as size of quantum dot decreases (from [13]).

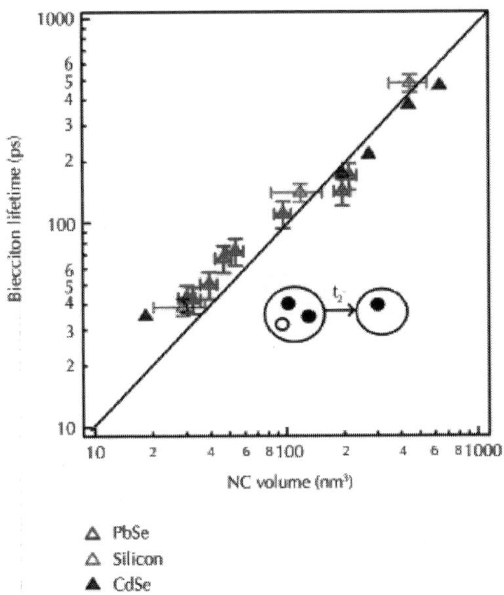

Figure 7: Biexction lifetime as a function of volume of nanocrystal (from [13]).

In [32], authors have reported excitation of seven electrons from a photon in PbSe nanocrystal and have reported an internal quantum efficiency of 700%. Based on the type of measurements reported in [32], one is tempted to believe that for solar cells more than one EHP could be generated for each incoming photon, which in turn will improve the solar cell efficiency. In reality, for solar cell efficiency enhancement these multiple EHPs must be extracted from the device through electrodes. Fundamentally this is not possible, since all the reported internal quantum efficiencies of above 100% are based on the measurement of Transient Absorption Spectra (TAS). In TAS, two ultrafast femtosecond LASER pulses are used. First, the high energy LASER beam is used to excite the electrons and the other LASER beam probes the excited states. The pumped power used for multiple exciton generation in [6] is 1.126 mW/cm^2 which is roughly an order of magnitude more than the power available at 600 nm in AM1.5 solar spectrum. The TAS measurements may give information about the excited states [28], but these experimental conditions have nothing to do with the required conditions for solar cell applications.

As early as 1980 one of us [33] provided experimental data to refute the claim that one can observe short circuit current density greater than predicted by Shockley and Queisser limit [25]. An extensive search of the open literature beyond 1980 did not show any direct current-voltage measurements under illumination or enhancement in efficiency of solar cells by utilizing hot carriers or by any other mechanism including generation of multiple excitons.

Up Conversion and Down Conversion.

Up/Down conversion involves altering the incident spectrum which is available to the solar cell. In down conversion, one high energy photon will be split in to two or more photons which can be converted into two or more electron hole pairs by the solar cell. In up conversion, two or more low energy photons are converted in to a single photon which can be converted by the solar cell to generate an electron hole pair. Figure 8 shows the operating mechanism of proposed down converter solar cell. The solar cell is a regular siliconbased solar cell with energy band gap of E_g . It converts only photons of energy greater than E_g in to EHPs. According to proposed down conversion mechanism, if a photon with energy of 2Eg strikes the solar cell and passes through it and further strikes the converter placed behind the solar cell which has an energy band gap of $2E_g$. This results in exciting an electron from the valence band to the conduction band. This excited electron falls back into the valence band in two steps. In the first step, it falls to an intermediate state emitting a photon "1" of energy E_g and then in the second step, it falls down in to the valence band emitting another photon "2" as shown in Figure 8 of energy E_g . These two photons with energy E_g will go back to the solar cell and get converted in to two EHPs. Thus, the single high energy photon will result in two EHPs which is not possible in a conventional solar cell. Also, the solar cell is assumed to be very thin to pass photon with energy $2E_g$ which could have a negative impact on the efficiency

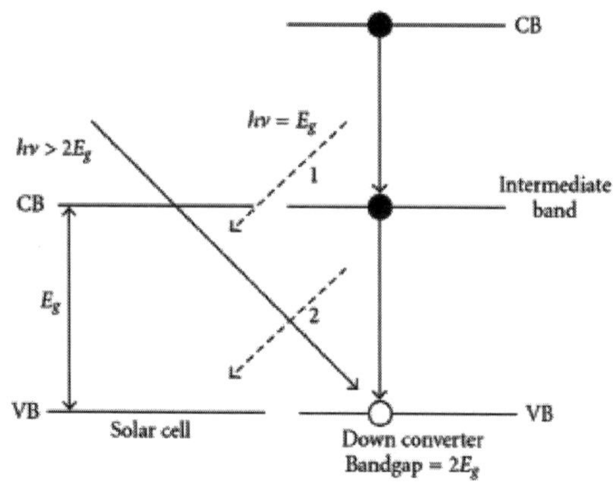

Figure 8: A photon with energy more than $2E_g$ strikes on down converter and produces two photons with energy E_g.

The theory of high efficiency down conversion process is proposed on the assumption that "no nonradiative recombination takes place anywhere in the device" [34]. For a photon of energy greater than $2E_g$, a band-toband transition happens in the converter and these excited electrons recombine via intermediate bands in the converter. All the photons with energy between E_g and $2E_g$ will be wasted if they cross the solar cell. Also, photons with energy more than $2E_g$ will be absorbed by the converter and will result in thermalization which will further decrease the efficiency. If the down converter is placed above the solar cell, it will absorb all photons with energy greater than its band gap. Thus, the solar cell collects photons only generated by the down converter via photoluminescence. Also, only half of the photons can be expected to be radiated into the solar cell. The other half will be radiated outwards from the convertor.

With the assumptions described above, the AM 1.5 efficiency limit for a down-conversion solar cell with $E_g = 1.05$ eV is 39.6% [34] while the Shockley and Queisser limit is about 33% for single junction silicon solar cell [25]. The mathematical calculations about up conversion also have fundamental flaws and the efficiency limit for this process is calculated as 47.6% for nonconcentrated sunlight [35]. The authors of [36] have conducted a review on the existing materials to modify the solar spectrum and a few results on solar cell up/down conversion using nanostructures were also reported. Most of the results reported in [36] are nothing other than the explorations of the photoluminescence properties of materials.

Use of Si nanocrystals (NCs) as a down converter is reported in [37] which show no change in open circuit voltage and a little increase in short circuit current which results in net efficiency increase by only 0.4%. Also, it is worth mentioning here that in [37] the results of only one sample were reported for claiming the efficiency improvements. As reported in [37], the difference between the value of current density measured by spectral response and the value extracted from electrical measurement data is 40%. Thus the legitimacy of the claim by authors of [37] about the 0.4 % net efficiency improvement by the use the down converter based on silicon nanocrystals is questionable. The internal quantum efficiency results reported in [37] are shown in Figure 9. The reduction in internal quantum efficiency for cells annealed at 200◦C and 500◦C indicates device reliability problems. [37] was published in 2004 and in the last five years no new results about improving the efficiency by up/down conversion have been reported.

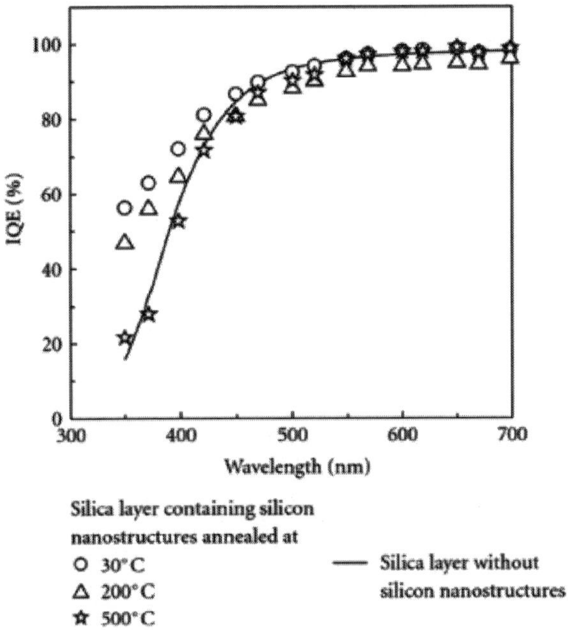

Figure 9: Measured internal quantum efficiency of solar cell with (symbols) and without a silica layer containing silicon nanocrystals. The reduction in internal quantum efficiency for cells annealed at 200° C and 500° C indicates reliability problem (from [36]).

In summary, no one has reported experimental results of the up/down conversion process with a meaningful efficiency enhancement to pursue these nanostructure-based solar cells for further research and development.

Nanocrystals-Based Tandem Solar Cells

As we stated earlier, the full solar spectrum cannot be efficiently absorbed by a solar cell using single energy band gap material. In 1955 for the first time, Jackson proposed to use layers of different energy band gap materials to enhance the solar cell efficiency [38]. III–V semiconductor and amorphous silicon-based tandem solar cells are already in production. A lot of research has been done in the search for a perfect combination of materials to absorb the entire solar spectrum. This motivated researchers to think for the application of low-dimensional structures such as quantum wells as an absorber layer in solar cells. The use of quantum wells in solar cells was first proposed by Burnham and Duggan in 1990 [39]. It is well-known fact that the energy band gap of quantum confined structures can be tuned by altering their size.

Very recently Cho et al. [40] have shown an efficiency of 10.6% by embedding 3 nm quantum dots in 2 nm SiO_2 layers. They used 15 layers of the phosphorous doped quantum dots embedded in SiO_2 on a p type crystalline silicon wafer. Also, they claimed that the current density of this cell is 29.8 mA cm−2 and it is equivalent to the current density of single junction nontextured solar cell [40]. Table 2 summarizes their experimental results. It can be seen from Thermal conductivity (W/mK) 20 40 60 80 100 120 140 160 Temperature (K) 250 300 350 400 450 500 550 SiGe QDS Bulk Si Figure 10: In plane lattice thermal conductivity of $Si_{1-\delta}Ge_{\delta}$ quantum dot superlattice and bulk silicon as a function of temperature. The value of x is 0.05 and size of quantum dot is 4 nm (from [43]). Table 2 that there is not any significant difference in the effi- ciency of various sized quantum dots embedded in the oxide layers. However, the maximum reported efficiency from their work is far below the efficiency of commercially available single junction silicon solar cells. In case of quantum dots as well as other nanostructures-based semiconductor devices fundamental issues related to heat removal and reliability are very important [24]. Therefore thermal conductivity of quantum dots [41–44] must be considered for applications of quantum dots in solar cells. As shown in Figure 10 [43], the thermal conductivity of the $Si_{1-\delta}Ge_{\delta}$ quantum dot super lattice is much lower than that of bulk silicon. The scattering of acoustic phonons on quantum dots is key scattering mechanism for decrease in the thermal conductivity [43]. Figure 11 shows the in-plane lattice thermal conductivity as a function of dot volume fraction (**δ**) of Germanium in $Si_{1-\delta}Ge_{\delta}$ alloy with dot size as the variable parameter. Additional complexity in the applications of quantum dots in solar cells will arise from the fact that quantum dots in the order of quadrillion (10^{15}) will be required for a silicon wafer of the size of 150 mm. Physics related to many body problems must be considered to understand the interactions among various quantum nanocrystals [45]. Control on the size distribution of nanocrystals is another challenge. The interaction between nanocrystals increases when size distribution of nanocrystals is large and results in reduced mobility and short circuit current. The reason for the mobility reduction is hopping of charge carriers between adjacent nanocrystals [17].

Table 2: Electrical measurements of solar cell with different size silicon nanocrystals embedded in layers of silicon rich oxide (SRO)/Silicon Dioxide (SiO_2) under standard condition (AM 1.5 spectrum, 100 mW cm^{-2}, 298° K) (from [40])

SRO/ SiO_2thickness,number of bilayers	Open circuitvoltage (mV)	Short circuitcurrent density j_{sc}(mA cm^{-2})	Fill factor(%)	Efficiency(%)
3 nm/2 nm 15 BL	555.6	29.8	63.8	10.6
4 nm/2 nm 25 BL	540.3	25.0	76.8	10.4
5 nm/2 nm 25 BL	517.9	27.9	72.3	10.5
8 nm/1 nm 25 BL	470.8	18.6	65.1	5.7

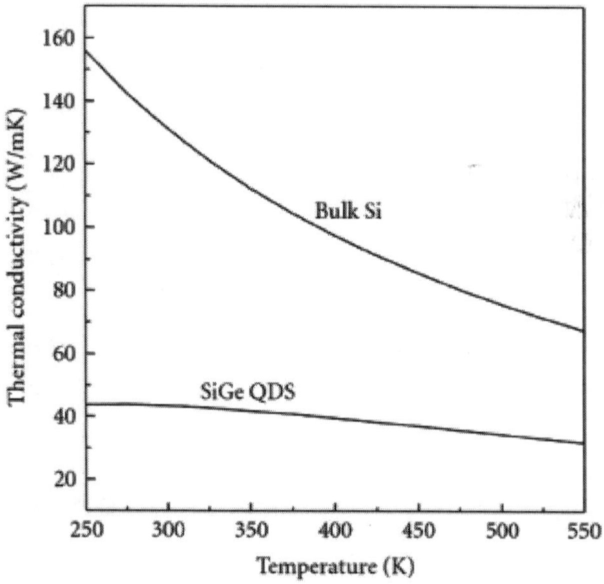

Figure 10: In plane lattice thermal conductivity of $Si_{1-\delta}Ge_{\delta}$ quantum dot superlattice and bulk silicon as a function of temperature. The value of x is 0.05 and size of quantum dot is 4 nm (from [43]).

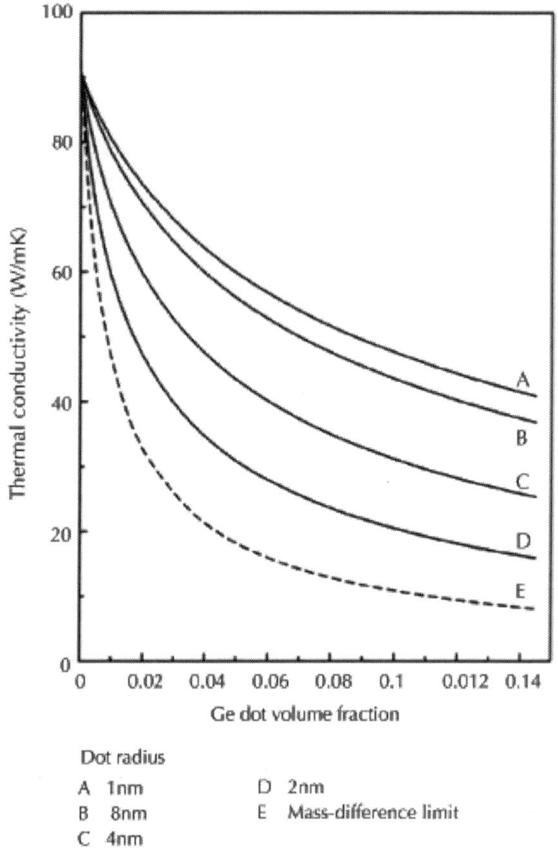

Figure 11: In-plane lattice thermal conductivity as a function of dot volume fraction (δ) of Germanium in Si1−δGeδ alloy. The value of dot size is 10.0 nm, 8.0 nm, 4.0 nm, and 2.0 nm for curve A, B, C, and D, respectively (from [43]).

Intermediate Band Solar Cells (IBSCs)

In 1960 Wolf described various loss factors in solar cells and discussed various mechanisms including multitransition solar cells (introducing energy levels in forbidden energy gap of active layer) to reduce energy losses [46]. Luque and Marti [47] introduced the concept of intermediate band solar cells in 1997 which is similar to the concept of multitransition solar cells. The basic concept of IBSC is to have an intermediate band to utilize low energy photons from the solar spectrum [47]. The formation of intermediate band is proposed by utilizing some impurities other than basic absorber material or alloy or quantum dots [48–50]. Fundamentally, the concept of IBSC is against the theory of recombination generation where energy levels near the mid-gap degrade the performance of all type of solar cells. This is indeed the case as has been confirmed by the experimental results [48]. The dark and light current density–voltage characteristics of GaAs reference cell, GaAs IBSC solar cells based on quantum dot with and without delta (δ) doping of quantum dot layers are shown in Figures 12 and 13, respectively. As expected the results of Figures 12 and 13 confirm that the introduction of intermediate band layer degrades the solar cells performance. Similar negative results about IBSC solar cells results have been reported by others [51].

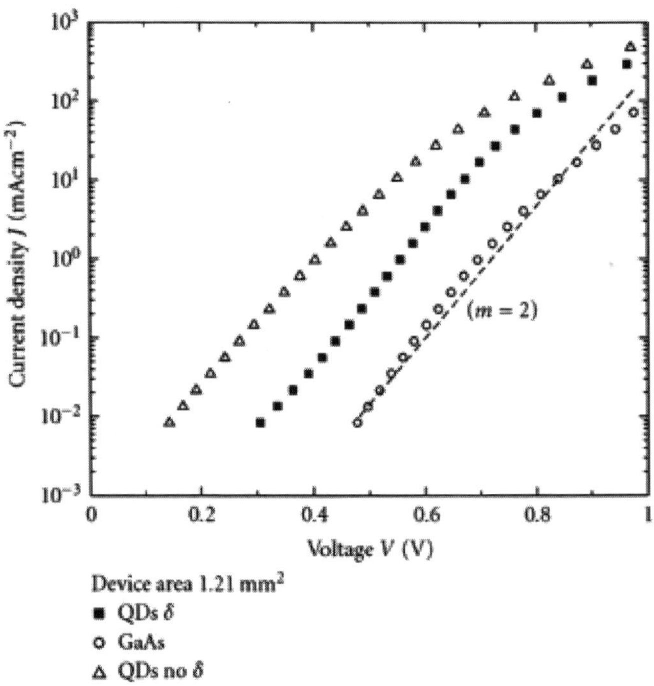

Figure 12: The dark current density-voltage characteristics of GaAs reference cell, GaAs IBSC solar cells based on quantum dot with and without delta (δ) doping of quantum dot layers (from [48]).

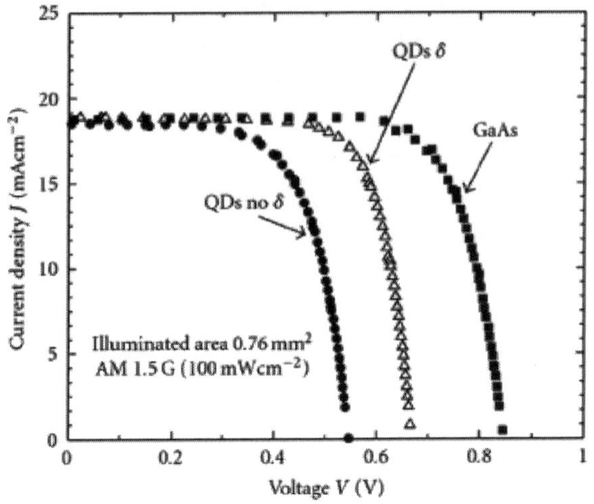

Figure 13: The light current density-voltage characteristics of GaAs reference cell, GaAs IBSC solar cells based on quantum dot with and without delta (δ) doping of quantum dot layers (from [48]).

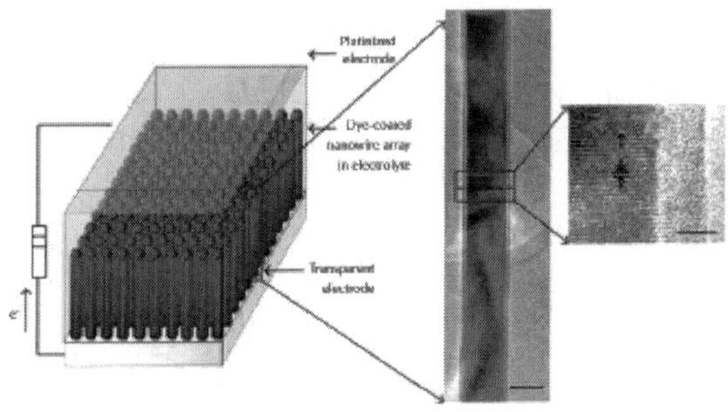

Figure 14: NW film-based DSC (from [54]).

Dye-Sensitized Solar

Cells. Dye-sensitized solar cells (DSCs) are currently being explored as ultra-low-cost solar cells. DSCs utilize three main components to perform photoelectro conversion which are metal oxide semiconductor nanostructures, dye, and electrolyte. A dye is used to create electron hole pairs and other two materials transport theses photogenerated charge carriers to the electrodes. The most popular metal oxide semiconductor nanostructures are made of Titania (TiO_2). Various researchers have proposed the use of nanotubes for DSCs and these results are discussed in the next paragraph. For a device of aperture area equal to 0.219 cm2 and platinum coated glass as the counter electrode, the best efficiency reported by researcher at Sharp for DSC is 11.18% [52]. Scaling to large area practical solar cells is a difficult challenge. A number of dyes including Ruthenium-based diketonato Ru (II) complex are used in DSCs. Ruthenium is a very costly metal and it requires extra recycling fee [11]. The hole transport material in DSCs, that is, electrolyte, can be a room temperature ionic liquid, quasisolid state, or solid state material [11]. A liquid electrolyte can fill the pores of the metal oxide semiconductor nanoparticles but it is not possible with a quasi-solid state or solid state electrolyte [18]. As a result, the best efficiencies are achieved for DSCs with liquid electrolyte. The liquid electrolytes get evaporated under exposure of solar radiations for longtime and poses fundamental problem of reliability.

Other than nanoparticles, nanowire (NW) film that provides a large surface area for the incorporation of light harvesting molecules, is being investigated to improve the performance of DSCs. However, NW DSCs rely on traplimited diffusion for electron transport, a slow mechanism that limits the device efficiency, especially at longer wavelengths. During operation, photons intercepted by the dye monolayer create excitons that are rapidly split at the NW surface, with electrons injected into the NW film and holes leaving the opposite side of the device by means of redox species in a liquid or solid-state electrolyte [53].

Another option for improving the absorption of red and near-infrared light is thickening the nanoparticle film to increase its optical density, which was unsuccessful because the film thickness comes to exceed the electron diffusion length through the nanoparticle network. Even incorporation of NW does not improve the efficiency. Solar cells constructed with ZnO NW arrays of various surface areas, at one full Sun intensity, have highest Jsc = 5.3–5.85 mA cm^{-2}, Voc = 0.61–0.71 V, FF = 0.36–0.38, and efficiency η = 1.2–1.5% [55]. Incorporation of NW (Figure 14) into the film halves the fill factors, limiting their potential efficiencies. Ideally, 20 μm long NW with a dominating diameter of 60 nm, a density of 10^{10} cm^{-2}, and surface area of 1080 cm^2per unit area of the substrate are needed where as the best that can be achieved is only 10% of ideal density and 18% of ideal surface area [54]. An effort to reduce the recombination rates through the use of core-sheath NW reduces the efficiencies to a maximum of 0.29% [56]. Titania shells 10–25 nm in thickness cause an increase in V_{OC} and fill factor but only with a current fall off, resulting in a overall conversion efficiency, only up to 2.25% under 100 mW cm^{-2} AM 1.5 simulated sunlight [57]. Raising the efficiency of NW DSCs above 2.5% depends on achieving higher dye loadings through an increase in NW array surface area, which is a non trivial challenge. Besides the lower efficiencies of these NW cells, we note that increasing the aspect ratio of the NW and achieving 90% higher densities are a daunting task

In summary, the reliability of DSC is a fundamental problem. In addition to reliability, a number of manufacturing issues that relate to scaling, materials, defect density, and cost of ownership need to be addressed. Without solving these fundamental problems, DSC will not play any significant role in manufacturing of next generation of PV modules.

Organic Solar Cells

Organic solar cells are solution processed PV devices and are attractive due to low processing temperature and simple fabrication techniques. After nearly three decades, it is still a challenge for organic solar cells to enter in the commercial market. Poor electrical and structural properties and fast degradation under solar radiation are the fundamental problems for current organic solar cells. The major material problems with organic solar cells are charge generation, geminate recombination, electric and magnetic field effects on charge separation and recombination, carrier mobility, polaronic effects, charge collection efficiency, organic semiconductors with optical gap in the near IR spectral range to better match the solar energy spectrum, interpenetrating morphologies, exciton range, and its dependence on the film morphology [58].

To date, the best efficiency achieved for organic solar cells using a tandem cell structure for solar intensity of 20 mW cm^{-2} and 200 mW cm^{-2} is 6.7% and 6.1%, respectively [59]. The device area was not mentioned in [59] and there may be a number of issues related to manufacturing of these devices. However, the fundamental problem is reliability. After 100 hours of continuous exposure under one sun (AM 1.5), the solar cell efficiency is degraded by 40 % [59].

In 2001 one of us [60] critically examined the prospects of manufacturing organic semiconductor-based integrated circuits and pointed that there is no appropriate technology that can be implemented for pilot-line production, eventually leading to mass-scale manufacturing. Fundamentally, the low thermal conductivity of polymer semiconductor is a serious problem. Low-cost polymer deposition techniques (e.g., coating, electrodeposition, etc.) have not provided the desired microstructure properties. With the present approaches under investigation, it is difficult to imagine the possibility of controlling the surface defects and voids which are detrimental to the performance device performance [60]. Similar arguments apply to organic solar cells. As we mentioned in Section 2.2 the low-cost film deposition techniques (e.g., roll to roll process) are quite attractive from capital expenditures. However due to high defect density, the cost of ownership of manufacturing for organic solar cells is high. Unless a new polymer is invented and manufacturing requirements discussed in Section 2 and reliability problems are solved, it is highly unlikely that organic solar cells will play any significant part in future manufacturing of photovoltaic modules.

DISCUSSIONS ON MANUFACTURABILITY OF NANO-STRUCTURE-BASED SOLAR CELLS

Figure 15 shows the evolution of a new concept leading to the mass-scale manufacturing of a new product. In case of nanostructure-based solar cells, currently we are in the very first block of Figure 15. The physics related to nanostructures mostly deals with quantum phenomena. There is nothing wrong with quantum physics or the associated quantum phenomena. The problem lies in stretching the results of observed phenomena beyond what is evident from the observed results. As an example in August 2007 Beard et al. [6] published a paper on multiple exciton generation in colloidal silicon nanocrystals and claimed that their findings are of particular importance for Si dominated photovoltaic solar cell industry. Based on the results of [6], Bullis published in

Technology Review [7] that by generating multiple electrons from high-energy photons, solar cells made of silicon nanocrystals could theoretically convert more than 40 percent of the energy in light into electrical power. As we have discussed in Section 3.1, there is no direct correlation between the photoluminescence results reported in [6] and the operation of solar cells. This type of hype exists both in technical journals as well as financial publications. Based on our in depth review of the literature we do not see any evidence that nanostructure-based solar cells will be part of PV module manufacturing. The lack of thorough understanding of the physical phenomena and the tendency of stretching observed experimental results is the fundamental reason that quantum phenomena-based nanostructures have a limited role to play in solar cells. Similar concerns about quantum phenomena-based nanostructures have been expressed for other semiconductor devices. According to Kelly [61], the new physics associated with one-dimensional and zero-dimensional structures, quantum interference, ballistic motion, or Coulomb blockade is unlikely to lead to practical devices until these effects can be exhibited, if ever they can, as robust phenomena.

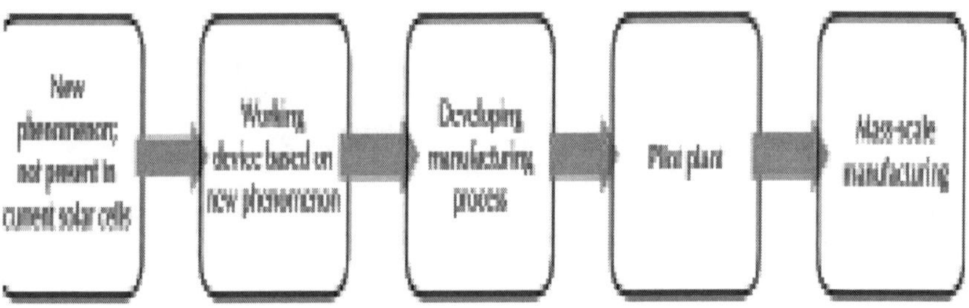

Figure 15: Basic product development and mass-scale manufacturing cycle.

Dye-sensitized solar cells (DSCs) and organic solar cells are currently being investigated by several companies. Other than manufacturing related issues, reliability is the main concern. It is worth mentioning here that the public and or private investment in a company does not mean that the investment will bring any product in the market and generate revenue. Due to lack of knowledge of manufacturing challenges, many times the investors think massscale manufacturing as a trivial exercise and the company is funded. Even for well-established companies, lack of necessary and sufficient data of defects, yield, and cost of ownership can result in failure. Based on the research work, Motorola formed a subsidiary called Thoughtbeam in November 2001 to develop, sell, and license its GaAson-silicon technology with applications to less-expensive optical communications, high-frequency radio devices, and high-speed microprocessor-based systems. Due to the fundamental nature of defects densities at the interface of GaAs and silicon and inability to deliver the desired device and circuit performance the subsidiary company was shut down in January 2003 [62]. Lack of manufacturing experience and not following the development and manufacturing cycle shown in Figure 15 can lead to wastage of millions of dollars. With an investment of at least $322 million, Optisolar has to close operation due to manufacturing problems [63].

Based on the technical and business issues discussed in this section, it is obvious that none of the proposed nanostructure based solar cells that violates laws of physics, have any prospects of manufacturing PV modules. Due to the issues related to materials availability and toxicity of materials discussed in Section 2, CdTe, CuInGeSe2, and III–V semiconductor-based solar cells will not become a dominant player in multibillion dollar PV market. As one of us predicted in 1980 [20], silicon-based PV manufacturing will provide sustained growth of the PV industry and will emerge as a low-cost viable source of electricity generation for the 21st century. Following are the four silicon PV technology that are currently used in manufacturing, (i) ultra high efficiency single crystal technology, (ii) medium efficiency poly or multicrystalline technology, (iii) amorphous silicon thin film technology based on the use of larger glass substrates (⊠5.2 m2), and (iv) amorphous silicon technology based on flexible substrates. From the low-cost and multiterawatt PV power generation point of view, only amorphous silicon solar cells based on the use of large glass substrates have the potential to reach the goal of providing installed PV system cost of $\$2.00/w_p$ in the next 3-4 years [64]. The ultra high efficiency single crystal PV modules may find a small niche market (where small area PV modules are warranted due to the area available for installation). The amorphous silicon technology based on flexible substrates may find a niche market for consumer products. The medium efficiency poly or multicrystalline silicon modules (limited size of silicon wafer) will not be able to be cost-effective compared to large area glass-based amorphous silicon thin film technology.

CONCLUSIONS

We have presented a comprehensive review on the subject of prospects of nanostructure-based solar cells for manufacturing future generation of photovoltaic modules. In search of reducing heat loss and utilizing the unabsorbed part of the spectrum, a number of solar cell concepts have been proposed. We have examined all proposed nanostructures for solar cell applications using the following five fundamental physics-based criterions.

- Interactions between charge carriers (electrons and holes) and phonon cannot be neglected.
- For all the known semiconductors, hot carriers (due to ultra short value of thermalization time) generated in one layer of the solar cell cannot be transported to the other side of the solar cell and collected at the other electrode.
- The phenomenon of photoluminescence has no direct relationship with solar cell operation.
- Intermediate band gap recombination center will reduce solar cell performance.
- For nanostructures, the thermal conductivity is lower than the bulk values. For practical applications of nanostructures, the material must have high thermal conductivity to transport heat from the device.

Most of the work reported in the literature is based on assumptions which are not realistic. Infinite carrier mobility, optimal absorption, and lossless contacts are some examples of unrealistic assumptions used in the work reported in the literature. With current knowledge base, it is highly unlikely that the nanostructure-based solar cells employing Hot Carriers, Up/Down conversion, and Intermediate Band Solar Cells will be used in the manufacturing of PV modules. Similar to the invention of phase shift masks (to beat the conventional diffraction limit of optical lithography); clever design concepts need to be invented to take advantage of quantum-based nanostructures.

Both dye-sensitized solar cells and organic solar cells have fundamental problem of reliability. Silicon-based PV manufacturing will continue to provide sustained growth of the PV industry and will emerge as a low-cost viable source of electricity generation for the 21st century.

ACKNOWLEDGMENTS

The authors would like to thank Elsevier, Wiley-VCH Verlag GmbH & Co. KGaA, Institute of Physics and American Institute of Physics for granting us permission to republish some of the figures and tables used in this paper. The authors would also like to thank Dr. M. Green, Dr. M.C. Beard, Dr. A. Khitun, Dr. A. Freundlich, Dr. P. Yang, and Dr. A. Martí for their quick response for our e-mail requests.

REFERENCES

1. M. Osborne, "Worldwide photovoltaics installations grew 110% in 2008, says Solarbuzz," March 2009,http://www.pv-tech.org/news/_a/worldwide_photovoltaics_installations_grew_110_in_2008_says_solarbuzz.
2. M. Osborne, "Gartner posts long-range forecast for photovoltaics industry," March 2009,http://www.pv-tech.org/news/_a/gartner_posts_long range_forecast_for_photovoltaics_industry.
3. R. Singh, N. Gupta, and K. F. Poole, "Global energy conversion revolution in the 21st century through solid state devices," in Proceedings of the 26th International Conference on Microelectronics (ICM '08), vol. 1, pp. 45–54, IEEE, Nis, Serbia, May 2008.
4. G. Conibeer, M. Green, R. Corkish, et al., "Silicon nanostructures for third generation photovoltaic solar cells," Thin Solid Films, vol. 511-512, pp. 654–662, 2006.
5. R. Singh, A. Venkateshan, K. F. Poole, D. Mohan, and P. Chatterjee, "Semiconductor manufacturing in the nanotechnology world of the 21st century," in Proceedings of the 25th International Conference on Microelectronics (ICM '06), pp. 2–8, Belgrade, Serbia, May 2006.

6. M. C. Beard, K. P. Knutsen, P. Yu, et al., "Multiple exciton generation in colloidal silicon nanocrystals,"Nano Letters, vol. 7, no. 8, pp. 2506–2512, 2007.
7. K. Bullis, "Nanocrystals for Superefficient Solar Cells," Technology Review, March 2007,http://www.technologyreview.com/energy/19256.
8. Department of Energy, Solar Energy Technologies Program, "Future Generation Photovoltaic Devices and Processes Selections," November 2007, http://www.docstoc.com/docs/942337/Future-Generation-Photovoltaic-Devices-and-Processes-Selections.
9. V. I. Klimov, "Mechanisms for photogeneration and recombination of multiexcitons in semiconductor nanocrystals: implications for lasing and solar energy conversion," Journal of Physical Chemistry B, vol. 110, no. 34, pp. 16827–16845, 2006.
10. A. Luque and A. Martí, "Ultra-high efficiency solar cells: the path for mass penetration of solar electricity," Electronics Letters, vol. 44, no. 16, pp. 943–945, 2008.
11. L. M. Goncalves, V. Z. Bermudez, H. A. Ribeiro, and A. M. Mendes, "Dye-sensitized solar cells: a safe bet for the future," Energy and Environmental Science, vol. 1, pp. 655–667, 2008.
12. A. J. Nozik, "Spectroscopy and hot electron relaxation dynamics in semiconductor quantum wells and quantum dots," Annual Review of Physical Chemistry, vol. 52, pp. 193–231, 2001.
13. M. C. Beard and R. J. Ellingson, "Multiple exciton generation in semiconductor nanocrystals: toward efficient solar energy conversion," Laser and Photonics Reviews, vol. 2, no. 5, pp. 377–399, 2008.
14. L. Tsakalakos, "Nanostructures for photovoltaics," Materials Science and Engineering R, vol. 62, no. 6, pp. 175–189, 2008.
15. A. Luque, A. Martí, and A. J. Nozik, "Solar cells based on quantum dots: multiple exciton generation and intermediate bands," MRS Bulletin, vol. 32, no. 3, pp. 236–241, 2007.
16. J.-F. Guillemoles and S. Kettemann, "Advanced concepts for photovoltaic conversion: bridging the gap from theory to real devices?" http://www.physnet.uni-hamburg.de/hp/kettemann/ispra.pdf.

17. V. V. Kislyuk and O. P. Dimitriev, "Nanorods and nanotubes for solar cells," Journal of Nanoscience and Nanotechnology, vol. 8, no. 1, pp. 131–148, 2008.
18. H. J. Snaith and L. Schmidt-Mende, "Advances in liquid-electrolyte and solid-state dye-sensitized solar cells," Advanced Materials, vol. 19, no. 20, pp. 3187–3200, 2007.
19. R. Licklider, "The power of oil: the Arab oil weapon and the Netherlands, the United Kingdom, Canada, Japan, and the United States," International Studies Quarterly, vol. 32, no. 2, pp. 205–226, 1988.
20. R. Singh and J. D. Leslie, "Economic requirements for new materials for solar photovoltaic cells," Solar Energy, vol. 24, no. 6, pp. 589–592, 1980.
21. A. Feltrin and A. Freundlich, "Material considerations for terawatt level deployment of photovoltaics,"Renewable Energy, vol. 33, no. 2, pp. 180–185, 2008.
22. H. Wicht and S. deHaan, "Hann, Photovoltaics Raw Materials Tracker," 2008,http://www.isuppli.com/Abstract/ABSTRACT%20-%20PV%20Raw%20Materials%20Market%20Tracker%202008.pdf.
23. S. Y. Sohn and H. U. Moon, "Cost of ownership model for inspection of multiple quality attributes,"IEEE Transactions on Semiconductor Manufacturing, vol. 16, no. 3, pp. 565–571, 2003.
24. R. Singh, P. Chandran, M. Grujicic, et al., "Dominance of silicon CMOS based semiconductor manufacturing beyond international technology roadmap and for many more decades to come," inSemiconductor Fabtech, pp. 104–113, 30th edition, 2006.
25. W. Shockley and H. J. Queisser, "Detailed balance limit of efficiency of p-n junction solar cells," Journal of Applied Physics, vol. 32, no. 3, pp. 510–519, 1961.
26. P. Würfel, A. S. Brown, T. E. Humphrey, and M. A. Green, "Particle conservation in the hot-carrier solar cell," Progress in Photovoltaics: Research and Applications, vol. 13, no. 4, pp. 277–285, 2005.

27. M. Wolf, R. Brendel, J. H. Werner, and H. J. Queisser, "Solar cell efficiency and carrier multiplication inSi1−xGex alloys," Journal of Applied Physics, vol. 83, no. 8, pp. 4213–4221, 1998.
28. Y. Takeda, T. Ito, R. Suzuki, T. Motohiro, S. Shrestha, and G. Conibeer, "Impact ionization and Auger recombination at high carrier temperature," Solar Energy Materials and Solar Cells, vol. 93, no. 6-7, pp. 797–802, 2009.
29. A. J. Nozik, "Multiple exciton generation in semiconductor quantum dots," Chemical Physics Letters, vol. 457, no. 1–3, pp. 3–11, 2008.
30. A. J. Nozik, "Quantum dot solar cells," Physica E, vol. 14, no. 1-2, pp. 115–120, 2002.
31. M. C. Hanna and A. J. Nozik, "Solar conversion efficiency of photovoltaic and photoelectrolysis cells with carrier multiplication absorbers," Journal of Applied Physics, vol. 100, no. 7, Article ID 074510, 2006.
32. R. D. Schaller, M. Sykora, J. M. Pietryga, and V. I. Klimov, "Seven excitons at a cost of one: redefining the limits for conversion efficiency of photons into charge carriers," Nano Letters, vol. 6, no. 3, pp. 424–429, 2006.
33. P. Smith, R. Singh, and J. Dubow, "Reverse current-voltage characteristics of indium tin oxide/silicon solar cells under illumination," Journal of Applied Physics, vol. 51, no. 4, pp. 2164–2166, 1980.
34. T. Trupke, M. A. Green, and P. Würfel, "Improving solar cell efficiencies by down-conversion of high-energy photons," Journal of Applied Physics, vol. 92, no. 3, pp. 1668–1674, 2002.
35. T. Trupke, M. A. Green, and P. Würfel, "Improving solar cell efficiencies by up-conversion of sub-band-gap light," Journal of Applied Physics, vol. 92, no. 7, pp. 4117–4122, 2002.
36. C. Strümpel, M. McCann, G. Beaucarne, et al., "Modifying the solar spectrum to enhance silicon solar cell efficiency—an overview of available materials," Solar Energy Materials and Solar Cells, vol. 91, no. 4, pp. 238–249, 2007.
37. V. Švrček, A. Slaoui, and J.-C. Muller, "Silicon nanocrystals as light converter for solar cells," Thin Solid Films, vol. 451-452, pp. 384–388, 2004.

38. E. D. Jackson, "Areas for improvement of the semiconductor solar energy converter," in Transactions of the Conference on the Use of Solar Energy, vol. 5, pp. 122–126, University of Arizona Press, Tucson, Ariz, USA, 1958.
39. K. W. J. Barnham and G. Duggan, "A new approach to high-efficiency multi-band-gap solar cells,"Journal of Applied Physics, vol. 67, no. 7, pp. 3490–3493, 1990.
40. E.-C. Cho, S. Park, X. Hao, et al., "Silicon quantum dot/crystalline silicon solar cells," Nanotechnology, vol. 19, no. 24, Article ID 245201, 2008.
41. B. O. Dabbousi, M. G. Bawendi, O. Onitsuka, and M. F. Rubner, "Electroluminescence from CdSe quantum-dot/polymer composites," Applied Physics Letters, vol. 66, pp. 1316–1318, 1995.
42. M. J. Konstantinovic, S. Bersier, X. Wang, et al., "Raman scattering in cluster-deposited nanogranular silicon films," Physical Review B, vol. 66, no. 16, Article ID 161311, 4 pages, 2002.
43. A. Khitun, A. Balandin, J. L. Liu, and K. L. Wang, "In-plane lattice thermal conductivity of a quantum-dot superlattice," Journal of Applied Physics, vol. 88, no. 2, pp. 696–699, 2000.
44. W. L. Liu, T. Borca-Tasciuc, G. Chen, J. L. Liu, and K. L. Wang, "Anisotropic thermal conductivity of Ge quantum-dot and symmetrically strained Si/Ge superlattices," Journal of Nanoscience and Nanotechnology, vol. 1, no. 1, pp. 39–42, 2001.
45. V. A. Belyakov, V. A. Burdov, R. Lockwood, and A. Meldrum, "Silicon nanocrystals: fundamental theory and implications for stimulated emission," Advances in Optical Technologies, vol. 2008, Article ID 279502, 32 pages, 2008.
46. M. Wolf, "Limitations and possibilities for improvement of photovoltaic solar energy converters—part I: considerations for earth›s surface operation," Proceedings of the IRE, vol. 48, pp. 1246–1262, 1960.
47. A. Luque and A. Martí, "Increasing the efficiency of ideal solar cells by photon induced transitions at intermediate levels," Physical Review Letters, vol. 78, no. 26, pp. 5014–5017, 1997.

48. A. Martí, N. López, E. Antolín, et al., "Novel semiconductor solar cell structures: the quantum dot intermediate band solar cell," Thin Solid Films, vol. 511-512, pp. 638–644, 2006.
49. K. M. Yu, W. Walukiewicz, J. Wu, et al., "Diluted II-VI oxide semiconductors with multiple band gaps,"Physical Review Letters, vol. 91, no. 24, Article ID 246403, 4 pages, 2003.
50. E. Antolín, A. Martí, J. Olea, et al., "Lifetime recovery in ultra highly titanium-doped silicon for the implementation of an intermediate band material," Applied Physics Letters, vol. 94, Article ID 042115, 2009.
51. A. G. Norman, M. C. Hanna, P. Dippo, et al., "InGaAs/GaAs QD superlattices: MOVPE growth, structural and optical characterization, and application in intermediate-band solar cells," in Proceedings of the 31st IEEE Photovoltaic Specialists Conference, pp. 43–48, 2005.
52. Y. Chiba, A. Islam, Y. Watanabe, R. Komiya, N. Koide, and L. Han, "Dye-sensitized solar cells with conversion efficiency of 11.1%," Japanese Journal of Applied Physics, vol. 45, no. 24–28, pp. L638–L640, 2006.
53. J. Krüger, R. Plass, M. Grätzel, P. J. Cameron, and L. M. Peter, "Charge transport and back reaction in solid-state dye-sensitized solar cells: a study using intensity-modulated photovoltage and photocurrent spectroscopy," Journal of Physical Chemistry B, vol. 107, no. 31, pp. 7536–7539, 2003.
54. P. Yang, "Nanowire-based solar cell," in Proceedings of the APS March Meeting, 2006, Session G10.00001.
55. M. Law, L. E. Greene, J. C. Johnson, R. Saykally, and P. Yang, "Nanowire dye-sensitized solar cells,"Nature Materials, vol. 4, no. 6, pp. 455–459, 2005.
56. L. E. Greene, M. Law, B. D. Yuhas, and P. Yang, "ZnO-TiO2 core-sheath nanorod/P3HT solar cell,"Journal of Physical Chemistry C, vol. 111, no. 50, pp. 18451–18456, 2007.
57. M. Law, L. E. Greene, A. Radenovic, T. Kuykendall, J. Liphardt, and P. Yang, "ZnO-Al2O3 and ZnO-TiO2 core-shell nanowire dye-sensitized solar cells," Journal of Physical Chemistry B, vol. 110, no. 45, pp. 22652–22663, 2006.

58. Z. V. Vardeny, A. J. Heeger, and A. Dodabalapur, "Fundamental research needs in organic electronic materials," Synthetic Metals, vol. 148, pp. 1–3, 2005.

59. Y. K. Jin, K. Lee, N. E. Coates, et al., "Efficient tandem polymer solar cells fabricated by all-solution processing," Science, vol. 317, no. 5835, pp. 222–225, 2007.

60. A. Lodha and R. Singh, "Prospects of manufacturing organic semiconductor-based integrated circuits,"IEEE Transactions on Semiconductor Manufacturing, vol. 14, no. 3, pp. 281–296, 2001.

61. M. J. Kelly, "Are quantum semiconductor devices delivering?" Materials Science and Engineering B, vol. 35, no. 1–3, pp. 1–6, 1995.

62. M. LaPedus and P. Clarke, "Motorola closes GaAs-on-silicon wafer subsidiary," EE Times, January 2003,http://www.eetimes.com/news/semi/showArticle.jhtml?articleID=10800321.

63. M. Kanellos, "Inside OptiSolar›s Grand Ambitions," April 2009,http://www.greentechmedia.com/articles/read/inside-optisolars-grand-ambitions-6029/.

64. R. Singh, unpublished results.

Chapter 4

Fabrication, Characterization, Properties, and Applications of Low-Dimensional BiFeO3 Nanostructures

Heng Wu, Jun Zhou, Lizhi Liang, Lei Li, and Xinhua Zhu

National Laboratory of Solid State Microstructures, School of Physics, Nanjing University, Nanjing 210093, China

ABSTRACT

Low-dimensional $BiFeO_3$ nanostructures (e.g., nanocrystals, nanowires, nanotubes, and nanoislands) have received considerable attention due to their novel size-dependent properties and outstanding multiferroic properties at room temperature. In recent years, much progress has been made both in fabrications and (microstructural, electrical, and magnetic) in characterizations of $BiFeO_3$ low-dimensional nanostructures. An overview of the state of art in $BiFeO_3$ low-dimensional nanostructures is presented. First, we review the fabrications of high-quality $BiFeO_3$ low-dimensional nanostructures via a variety of techniques, and then the structural characterizations and physical properties of the $BiFeO_3$ low-dimensional nanostructures are summarized. Their potential applications in the next-generation magnetoelectric random access memories and photovoltaic devices are also discussed. Finally, we conclude this review by providing our perspectives to the future researches of $BiFeO_3$ low-dimensional nanostructures and some key problems are also outlined.

INTRODUCTION

Multiferroics are formally defined as materials that exhibit more than one primary ferroic order parameter simultaneously [1, 2]. The coupling of different order parameters such as magnetoelectric coupling reveals the mutual regulation of magnetic and electric field which has potentially enormous commercial value in the next generation of information technology areas (e.g., multistate storage) [3–7]. Among the single-phase multiferroics, $BiFeO_3$ (BFO) has high Curie temperature (T_C=850°C), high Neel temperature (T_N=370°C), and large residual polarization intensity (150 μC/cm^2) at room temperature, which is widely investigated as model system of single-phase multiferroics. However, as one kind of G-type antiferromagnetic material BFO has spiral spin structure with a periodicity of 62 nm, this weak liner magnetoelectirc coupling makes it hard to be used in multifunctional devices [8]. Recently, it is found that the BFO nanoparticles exhibit relatively strong ferromagnetism as their sizes are below 62 nm [9]. Therefore, an enhanced magnetoelectric coupling can be achieved in the nanosized BFO materials, which play an important role in microelectronic devices [10–12]. Recent advances in science and technology of semiconductor integrated circuit have resulted in the feature sizes of microelectronic devices downscaling into nanoscale dimensions. At nanoscale the BFO materials display novel physical properties that are different from their bulk and film counterparts. Understanding the size effects of the BFO materials at nanoscale is of importance for developing the next generation of revolutionary electronic nanodevices. Due to the size effects being dependent on the structure and finite size, considerable efforts have been made in the controllable synthesis of low-dimensional BFO nanostructures such as nanowires, nanotubes, and their arrays. Much progress has also been achieved in the structure and property characterization. Furthermore, BFO nanostructures are also received extensive attention in the study of heterostructures [13] and domain characterizations [14–16]. This paper provides an overview of recent advances on the fabrication, structural characterization, and physical properties of low-dimensional BFO nanosized materials. Their potential applications are also discussed and some problems that need to be solved in future researches are also pointed out.

FABRICATION OF LOW-DIMENSIONAL BFO NANOSTRUCTURES

Generally, nanostructured materials can be classified into four classes by their dimensions, which are zero-dimensioned (0D) structures such as nanoparticles, one-dimensioned (1D) materials such as nanowires, nanorods, and nanotubes, two-dimensioned (2D) materials such as thin film and nanoislands, and three-dimensioned (3D) materials such as nanowires/tubes arrays, respectively. Up to date, two different approaches have been developed to fabricate nanostructured BFO, one is the bottom-up approach which means synthesizing nanostructures from atoms or molecules by assembling tiny into large, the other one is top-down approach which means dividing, etching or carving thin film or bulk materials into nanostructures as cutting big into small. In this section, we introduce the most used techniques for fabrication of low-dimensional BFO nanostructures and discuss their differences.

Sol-Gel Process

Sol-gel process is a kind of bottom-up approach, which generally involves the use of metal alkoxides and undergoes hydrolysis and condensation polymerization reactions to produce gels. After annealing the gels transform the porous gels into a dense target product.The main processing factors of sol-gel process are the water ratio in solution, pH value of solution, and its temperature. To synthesize BFO nanoparticles by sol-gel process, $Bi(NO_3)_3$ $5H_2O$ and $Fe(NO_3)_3$ $9H_2O$ are normally used as the raw materials. For example, Kim et al. [17] synthesized the BFO powders (average sizes ⊠200 nm) by dissolving $Bi(NO_3)_3$ $5H_2O$ into the mixture of 2-methoxyethanol and acetic acid, and as the solution became transparent they dissolved $Fe(NO_3)_3$ $9H_2O$ in it and kept the mixture at room temperature with stirring. Then, the precursor solution was dried at 80⊠ C for about 12 h to obtain the BFO xerogel powder. They grinded the xerogel powders and annealed them at 600⊠ C in air or N2 atmosphere to obtain the BFO nanosized powders. Gao et al. [18] dissolved the $Bi(NO_3)_3$ $5H_2O$ and $Fe(NO_3)_3$ $9H_2O$ with stoichiometric proportions into 2-methoxyethanol and adjusted the solution pH value to 4-5 by adding nitric acid subsequently. Then citric acid in 1:1 molar ratio with the metal nitrates and polyethylene glycol as a dispersant was added into the solution, respectively. After stirring the solution at 50⊠ C for 0.5 h, the solution was kept at 80⊠ C for 4 days to obtain the dried gel. After calcination at 500⊠ C for 2 h perovskite-type BFO nanoparticles with average diameter of ⊠80–120 nm were obtained. Similarly, Park et al. [9] also employed the sol-gel process to synthesize single-crystalline BFO nanoparticles and tuned their sizes from less than 15 nm to ⊠100 nm by changing the annealing temperatures. Obviously, the sizes of BFO nanoparticle were increased with the annealing temperature. However, the ideal synthetic pathway for the BFO nanoparticles is to obtain the appropriate size, shape, and crystallinity in the absence of the additional postannealing steps. Therefore, modified sol-gel techniques are developed to synthesize the BFO nanoparticles. Among them, hydrothermal method is one of the examples, in which there is no necessity for high-temperature calcination.

Hydrothermal Method

Hydrothermal method is also called the autoclave method, which involves heating an aqueous suspension of insoluble salts in an autoclave at a moderate temperature and pressure so that the crystallization of a desired phase will take place. The hydrothermal method is very popular method for synthesizing the perovskite nanoparticles because the synergetic effects from solvent, temperature, and pressure can offer stable final products and prevent the formation of impurity phases. The hydrothermal method is used to synthesize the BFO nanoparticles. For example, Chen et al. [19] utilized $Bi(NO_3)_3$ $5H_2O$ and $Fe(NO_3)_3$ $9H_2O$ as raw materials and adjusted the concentration of KOH at the region 1 M to 9 M to synthesize BFO nanoparticles with different morphologies at 200⊠ C for 6 h. Han et al. [20] used $Bi(NO_3)_3$ $5H_2O$, $Fe(NO_3)_3$ ⊠ $9H_2O$ and 8 M KOH to synthesize pure BFO powders at the temperatures of 175–225⊠ C and the hold time of 6 h, respectively. Wang et al. [21] used $Bi(NO_3)_3$ $5H_2O$ and $Fe(NO_3)_3$ $9H_2O$ as raw materials and synthesized the different morphologies of BFO nanoparticles by adding KNO3 mineralizer or not. All the above results demonstrate that the concentration of KOH determines the phase structure of the synthesized products. The pure BFO nanoparticles with diameters of 100–300 nm were synthesized when the concentration of KOH was 7 M and 12 M. If the concentration of KOH was below, the synthesized products were mainly composed of perovskite-type BFO accompanied with impurity phase of orthorhombic-type $Bi_2Fe_4O_9$. In hydrothermal procedure, the growth rate of $BiFeO_3$ was decreased by adding the KNO3 mineralizer, and thin plate-like $BiFeO_3$ could be obtained as increasing the reaction time. In addition, the solvothermal method is also a popular method to synthesize the BFO nanomaterials, which is similar to the hydrothermal method. The only difference is that the precursor solution for a solvothermal method is usually an organic medium such as ethanol or acetone, while that for a hydrothermal method is usually an aqueous medium such as water. Liu et al. [22] employed solvothermal method to synthesize single-crystalline BFO nanowires (45–200 nm in diameter, several nanometers to several micrometers in length) in acetone solvent. They dissolved $Bi(NO_3)_3$ $5H_2O$ and $Fe(NO_3)_3$ ⊠$9H_2O$ in 1 : 1 molar ratio then added deionized water and stronger ammonia water to adjust pH value of the solution to 10-11. After washing the sediment by distilled water until it is neutral, 5 M NaOH was added and kept stirring for 30 min. Then they heated the solution at 180⊠ C for 72 h hand obtained the single-crystalline BFO nanowires. Moreover, the morphology of the final products derived from solvothermal method can be effectively controlled by selecting the types of organics and their amount. For example, Zhang et al. [23] synthesized BFO nanoparticles and nanowires assembled by nanoparticles via changing the amount of the PVP or PEG (polyethylene glycol) polymer.

Template Method

Template method is based on chemical self-organization to synthesize nanostructured materials by the assistance of template which is porous with nanosized pores. By electrochemical deposition, sol-gel or chemical vapor deposition technology, atoms, or ions are deposited on the tube walls and form the nanostructure. Template method can be employed to fabricate one-dimensioned materials which have well dispersion and are easy to adjust aspect ratio of nanowires, tubes, and rods by changing the structural parameters of the used templates. Since the nanostructures fabricated by template method have similar structural properties (e.g., sizes, shape) to the templates, therefore, the pore sizes and their aspect ratios of the templates are crucial parameters for controlling the dimensions of the final nanostructures. Park et al. [24] first reported on the synthesis of BFO nanotubes by using porous anodic alumina (AAO) templates with pore-size of 100 or 200 nm. They dissolved $Bi(NO_3)_3$ $5H_2O$ and $Fe(NO_3)_3$ $9H_2O$ with a molar ratio of 1 : 1 to ethylene glycol and deposited the sol onto the surface of the AAO template by using a syringe. Finally they obtained BFO nanotubes with outer diameters in the range of 240 to 300 nm (for 200 nm pore-sized template) or 140 to 180 nm (for 100 nm pore-sized template) and lengths ranging from several microns to as much as 50 μm. However, the synthesized BFO nanotubes are polycrystalline because of the random nucleation on the walls of pores. Zhang et al. [25] also employed the similar method to synthesize BFO nanowires array viainjecting the sol on the template and annealed it subsequently. Gao et al. [26] synthesized polycrystalline BFO nanowires with a diameter of 50 nm and 5 μm in length by the similar method.

Synthesis of the BFO nanostructures by template method provides some advantages such as the structure of the nanoarray subject to the structure of the template, the channels of the template controlling the dimension of the materials, template pore walls preventing the aggregation of the material, and a large amount of nanowires or nanotubes mass produced.

Focus Ion Beam (FIB) Milling Method

The common top-down approach to fabricate nanostructured materials is focused ion beam (FIB) milling method. The typical technology is to utilize focused Ga^+ ions to bombard the thin film or bulk material to obtain the designed nanostructures. The advantages of FIB milling method are controllable morphology of structure or even patterns, facile operation and, and so forth. However, the disadvantages are high cost and the resolution of FIB is micron or submicron scale which is larger than the nanostructure.

Recently, Morelli et al. [27] fabricated BFO nanoislands by template-assisted FIB method based on epitaxial BFO thin films grown on $SrTiO_3$ (100) substrates. Arrays of 45 nm-thick aluminium dots were first evaporated on BFO thin films through template with aperture diameter of 400 nm. A focused ion beam with gallium ions was used to mill the specimen covered by Al dots. Chemical etching of the remaining Al was performed in 10% aqueous solution of potassium hydroxide (KOH) at room temperature for 90 s. Arrays of epitaxial BFO nanoislands with diameter ~250 nm were obtained. The features of top-down approaches are their precisely positioning and controlling the shapes and sizes of the designed BFO nanostructures. However, their time-consuming and low-throughput characters of these processes are the shortcomings of the top-down approaches.

Table 1 summarizes various approaches for fabricating the BFO low-dimensional nanostructures.

Table 1: Summary of various approaches for fabricating the BFO low-dimensional nanostructures

Approaches	Advantages	Disadvantages	Structures
Bottom-up			
Sol-gel	Facile regulation Low reaction temperature	Long reaction time	Nanoparticle
Template	Easy process uniform size and array	Limited aspect ratio	Nanowire/tube

Hydrothermal	Easy process high crystallinity low reaction temperature	Harsh reaction condition	Nanoparticle/wire
Top-down			
FIB	High resolution uniform size and array	Surface damage low throughput	Nanoisland/array

CHARACTERIZATION OF BFO NANOSTRUCTURES

Up to date various approaches including X-ray diffraction (XRD), scanning electron microscopy (SEM), (high-resolution) scanning transmission electron microscopy (HR) STEM, as well as X-ray energy dispersive spectrum (EDS), electron energy loss spectra (EELS), and selected area electron diffraction (SAED) have been developed to probe both the macroscopic and the microscopic details of BFO low-dimensional nanostructures. In this section, we will briefly summarize the recent the atomic-scale microstructural features of BFO low-dimensional nanostructures revealed a number of techniques.

BFO Nanoparticles/Nanoislands

Park et al. [9] first reported the synthesis of pure crystalline BFO nanoparticles by sol-gel method. Figure 1shows the structural characterizations of BFO nanoparticles with the average diameter of 95 nm by techniques of TEM, SAED, EDS, and HRTEM. Zhu et al. [28] also reported on the microwave-hydrothermal synthesis of spherical BFO nanoparticles, their TEM and HRTEM images are shown in Figure

2. The HRTEM pattern demonstrates the well crystallization of as-prepared BFO nanocrystals.

(a)

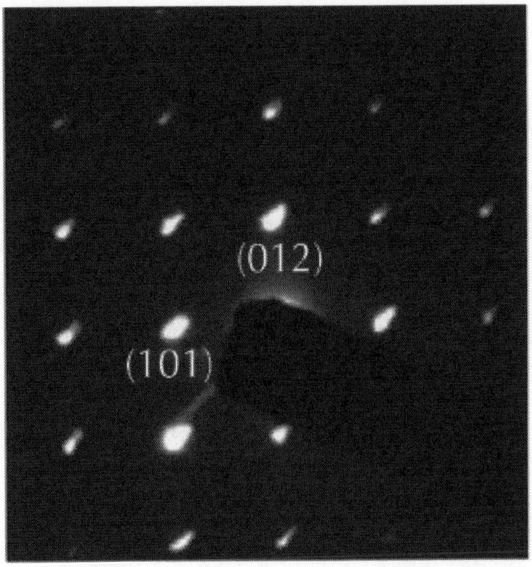

(b)

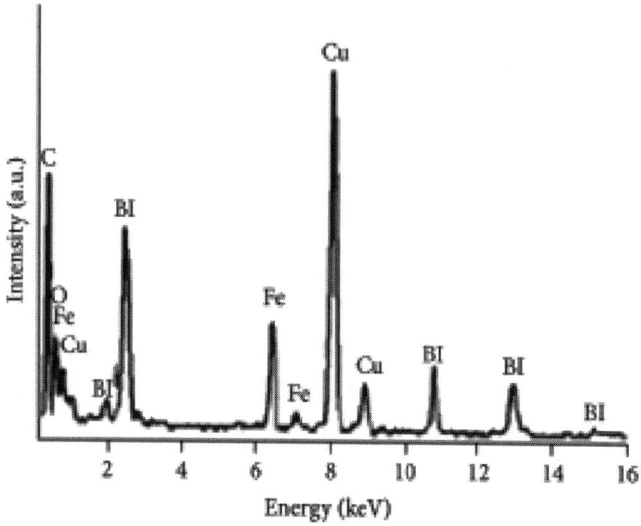

(c)

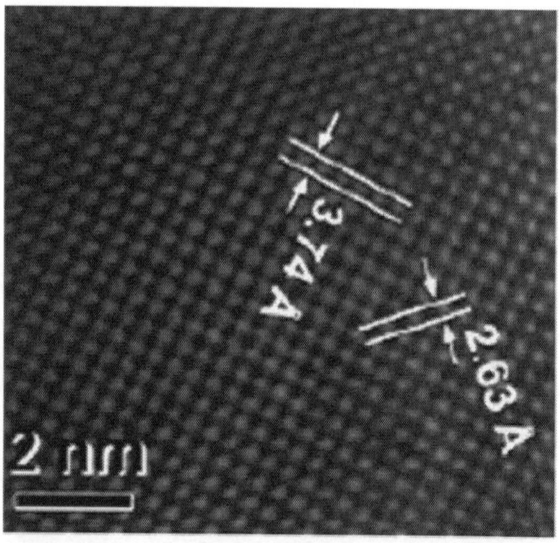

(d)

Figure 1: (a) TEM image, (b) SAED pattern, (c) EDS, and (d) HRTEM image of an individual $BiFeO_3$ nanoparticle (with a diameter of 95 nm) synthesized by sol-gel method.

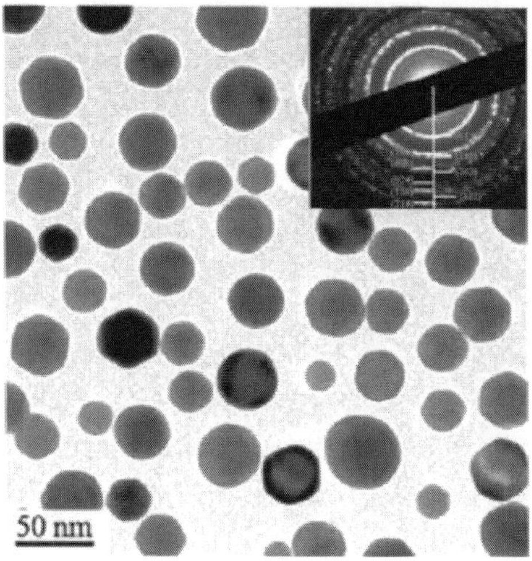

(a)

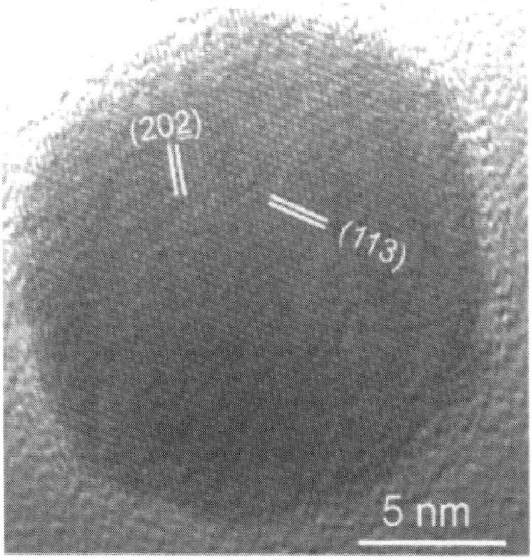

(b)

Figure 2: (a) TEM image of $BiFeO_3$ nanocrystals and the inset is selected area electron diffraction pattern, (b) HRTEM image of a single $BiFeO_3$ nanocrystal with a diameter of ~12 nm.

Besides the BFO nanoparticles, epitaxial BFO nanoislands were also synthesized by chemical self-assembled method [29], their phase structure and morphology were characterized by XRD and atomic force microscopy (AFM). Figure 3 shows the XRD patterns of the BFO nanoislands annealed at different temperatures; there is almost no other impurity peaks except the diffraction peaks of single crystalline $SrTiO_3$ substrate and the (100) crystal orientation of BFO. The AFM images shown in Figure 4 revealed that with increasing the postannealing temperature from 600°C to 800°C, the morphology of BFO nanoislands in the (100) growth plane evolved from triangled to squared, and then to plated shapes. Fractal ferroelectric domains and self-bias polarization were also found in a single BFO nanoisland, which were revealed by piezoforce microscopy (PFM) images.

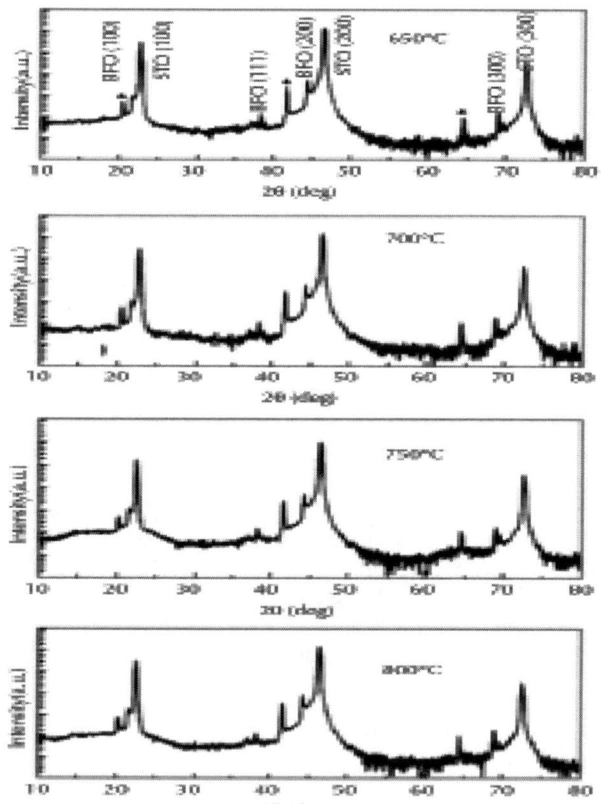

Figure 3: XRD patterns of the epitaxial $BiFeO_3$ nanoislands fabricated on $SrTiO_3$ (100) single crystal substrates by chemical assembled method, and

postannealed at 650°C; 700°C; 750°C; and 800°C for 1 hour. The diffraction peaks labeled by ▲ were from the $SrTiO_3$ (100) single crystal substrates diffracted by the Cu-K line due to the remaining Cu-K radiation.

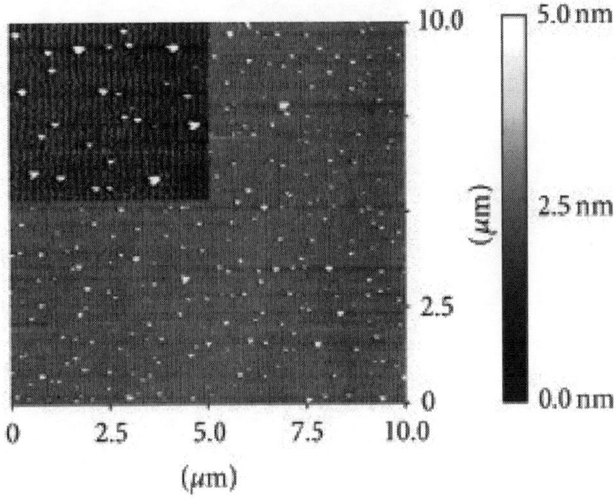

(a)

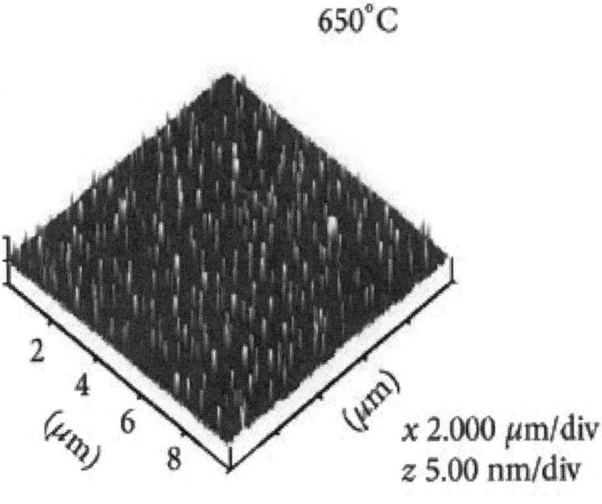

(b)

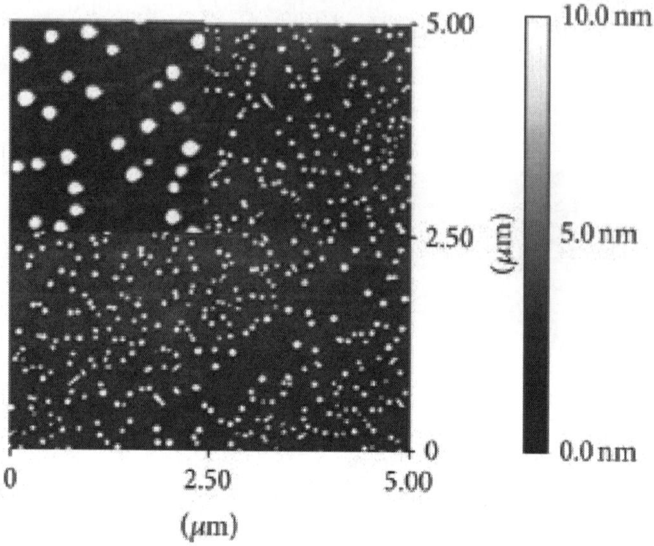
5.00
2.50
0
(μm)
0
2.50
5.00
(μm)
10.0 nm
5.0 nm
0.0 nm

(c)

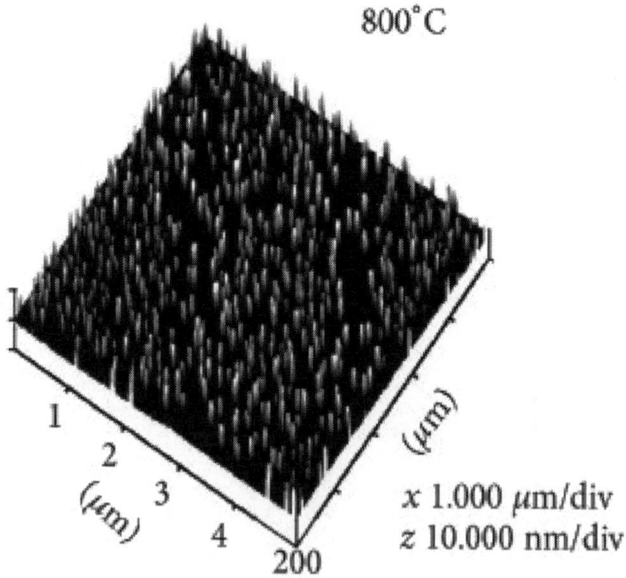
800°C
1
2
3
4
(μm)
(μm)
200
x 1.000 μm/div
z 10.000 nm/div

(d)

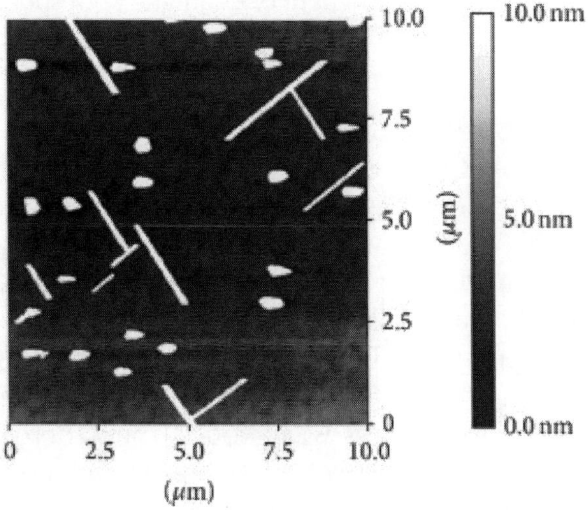

(e)

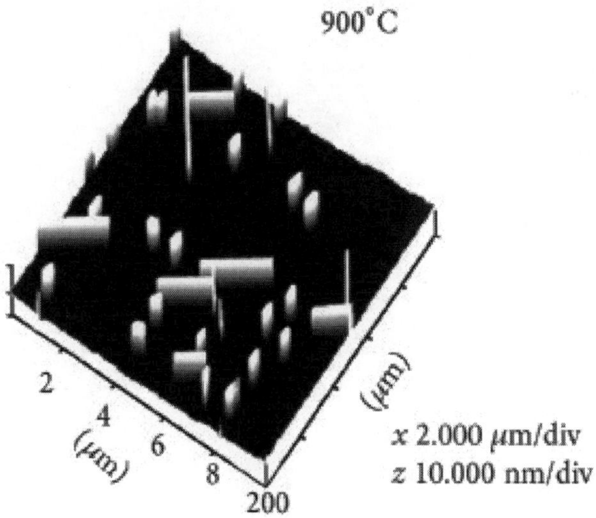

(f)

Figure 4: AFM images of the epitaxial $BiFeO_3$ nanoislands postannealed at ((a), (b)) 650°C, ((c), (d)): 800°C, ((e), (f)): 900°C for 1 h. The left AFM images are two-dimensional ones and the right ones are the corresponding 3-dimensional ones. The insets in (a) and (c) are the enlarged AFM images of the local surface areas.

Zhou et al. [30] also fabricated the BFO nanoring structure by combing sol-gel, AAO template-assisted, and planar TEM sample preparation methodology. Figure 5(a) shows the STEM image of the nanorings, where the atoms with high atomic numbers (such as Bi, Fe) exhibit bright image contrast; therefore, the BFO nanorings demonstrate high white bright contrast. Figure 5(b) shows the line scan of the intensity distribution of the STEM image contrast of the BFO nanorings, from which the inner diameter of the nanoring was determined to be about 170 nm and its thickness was about 20 nm. Therefore, the BFO nanorings were formed in the walls of the AAO template. The EDS data reveal that the BFO nanorings are composed of Bi, Fe, and O elements and that the chemical composition of the nanorings is close to $BiFeO_3$.

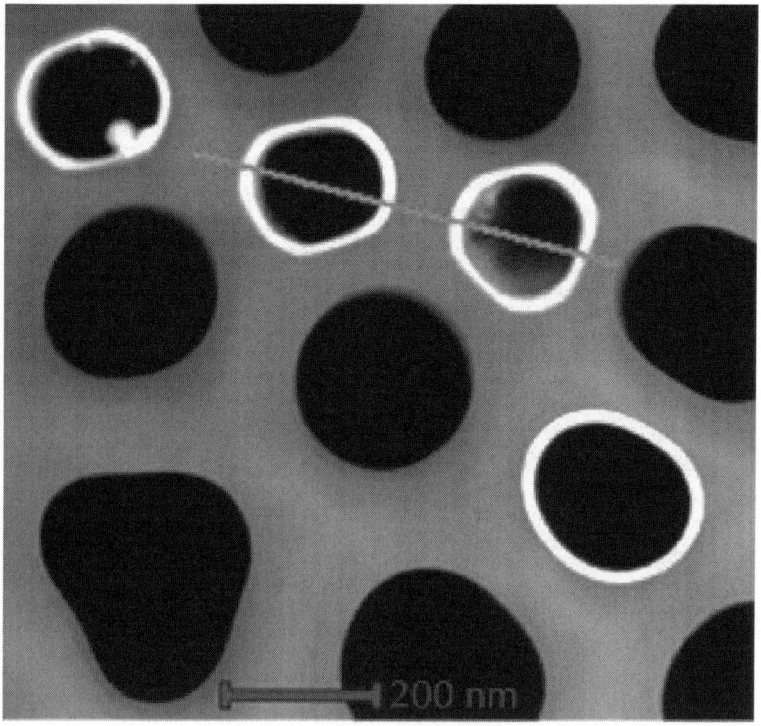

(a)

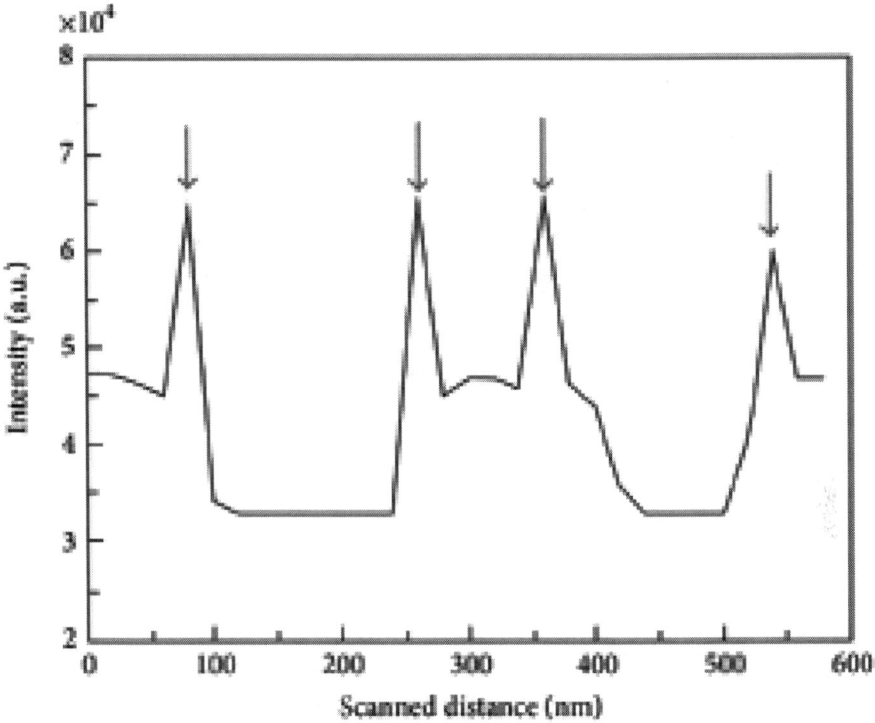

(b)

Figure 5: (a) STEM image of the $BiFeO_3$ nanorings, (b) line scan of the intensity distribution of the STEM image contrast of the $BiFeO_3$ nanorings (the scanned line indicated in (a)).

One-Dimensional BFO Nanostructures and Their Arrays

Zhang et al. [25, 31] fabricated the BFO nanowire and nanotube arrays by a template synthesis involving the sol-gel technique. Figure 6 shows the SEM images of the BFO nanowire and nanotube arrays, respectively. The TEM image of a single BFO nanotube is shown in Figure 6(c); the corresponding SAED pattern and the EDX spectrum are shown as insets.

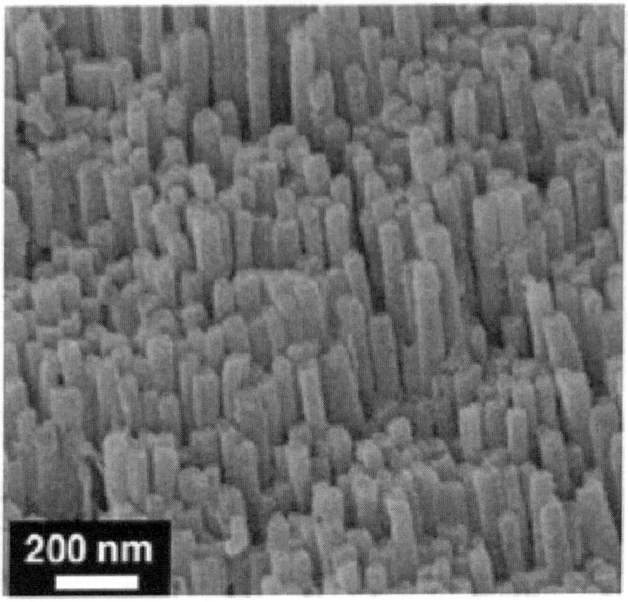

(a)

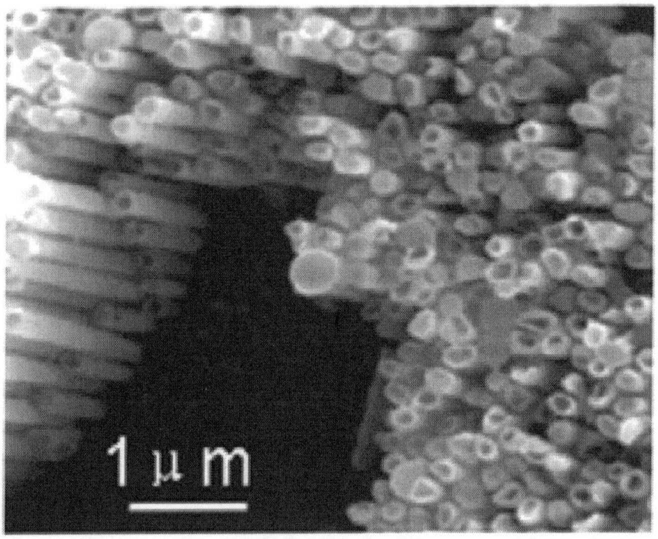

(b)

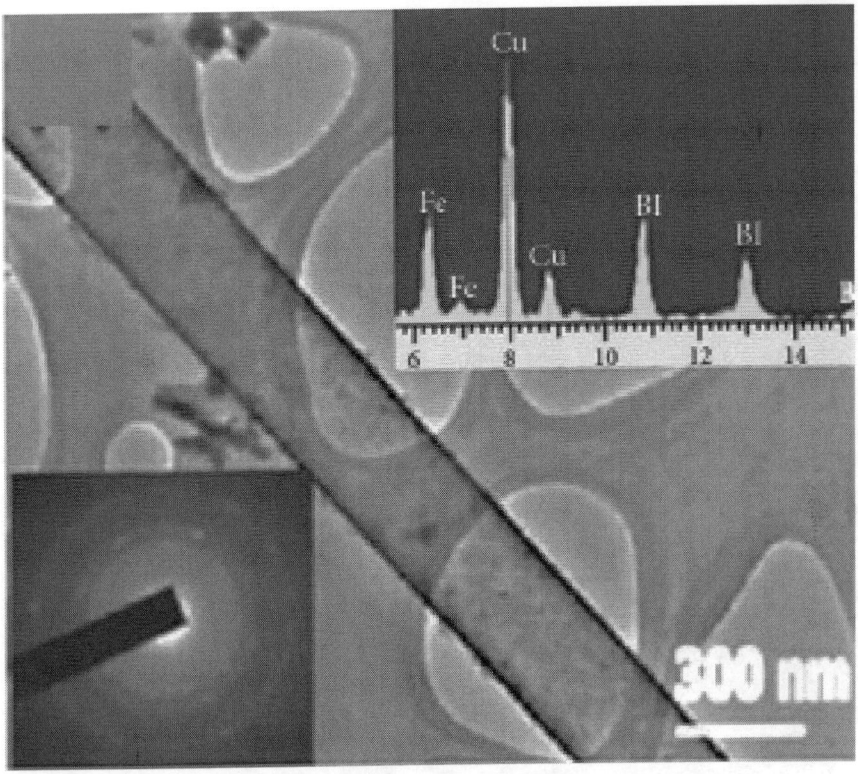

(c)

Figure 6: (a) SEM image of the $BiFeO_3$ nanowire array, (b) top-view of the nanochannel porous alumina filled with $BiFeO_3$ nanotubes, (c) TEM image of an isolated $BiFeO_3$nanotube, the left inset shows the corresponding SAED pattern and the right inset shows the EDX spectrum.

Domain Structures of BFO Thin Films

Chu et al. [32] characterized the ferroelectric polarization direction and domain structures of BFO films with different thicknesses (120, 15, and 2 nm) grown on $SrTiO_3$ (001) (top) and $DySrO_3$ (110) substrates (bottom) by PFM. Due to the competition between the normal strain caused by lattice mismatch and shear strain caused by the rhombohedral symmetry, ferroelectric polarization direction and domain structures are changed with the decrease of the film thickness, as shown in Figure

7. With decreasing the thickness of BFO films deposited on $SrTiO_3$ (001), the domain morphology evolved from stripe domain structures (Figure 7(a)) to intricate domain structures with fluctuate mottled contrast (Figures 7(b) and 7(c)). However, the domain size became larger as decreasing the thickness of BFO films deposited on $DySrO_3$ (110) substrates. Such a difference arises from the normal strain since it cannot be released at such thicknesses; therefore, BFO films on $SrTiO_3$ have larger elastic strain energy, which leads to smaller domain structures.

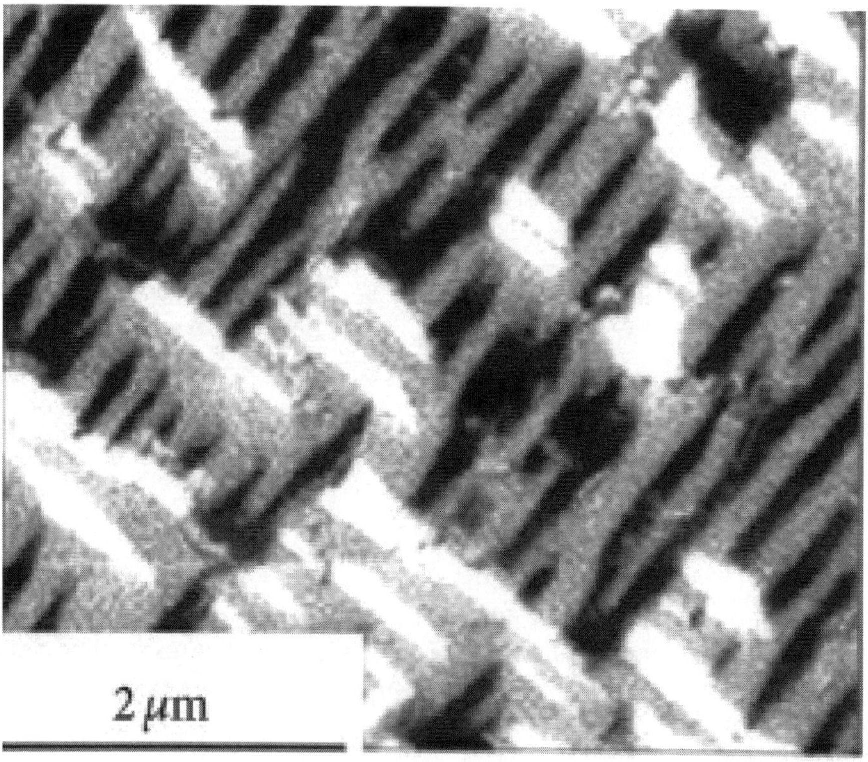

(a)

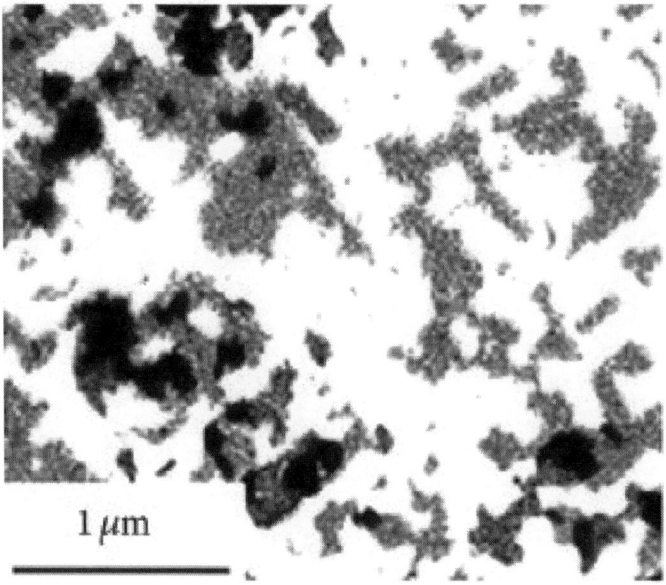

(b)

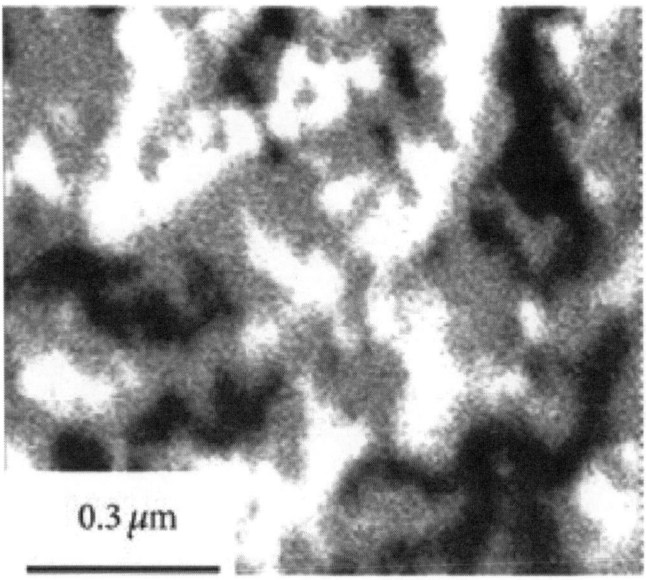

(c)

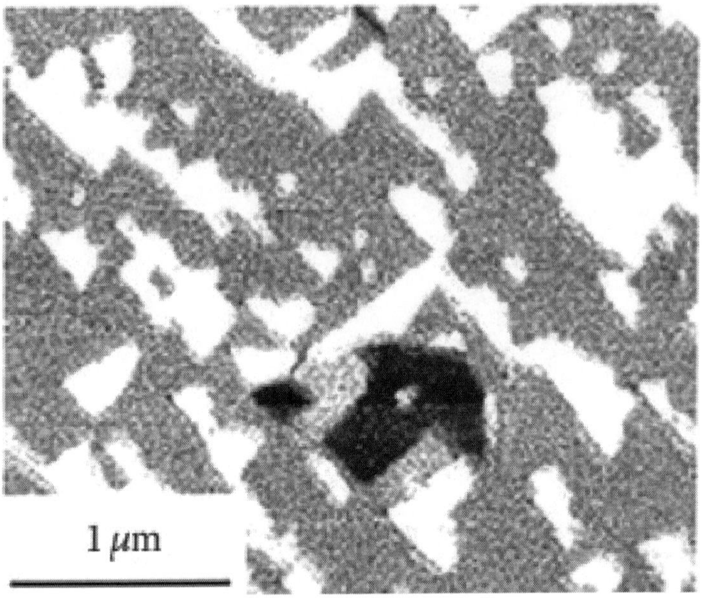

(d)

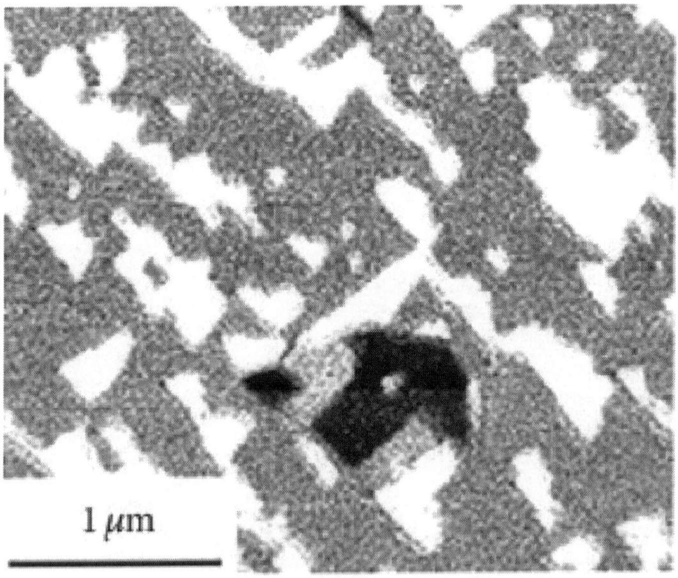

(e)

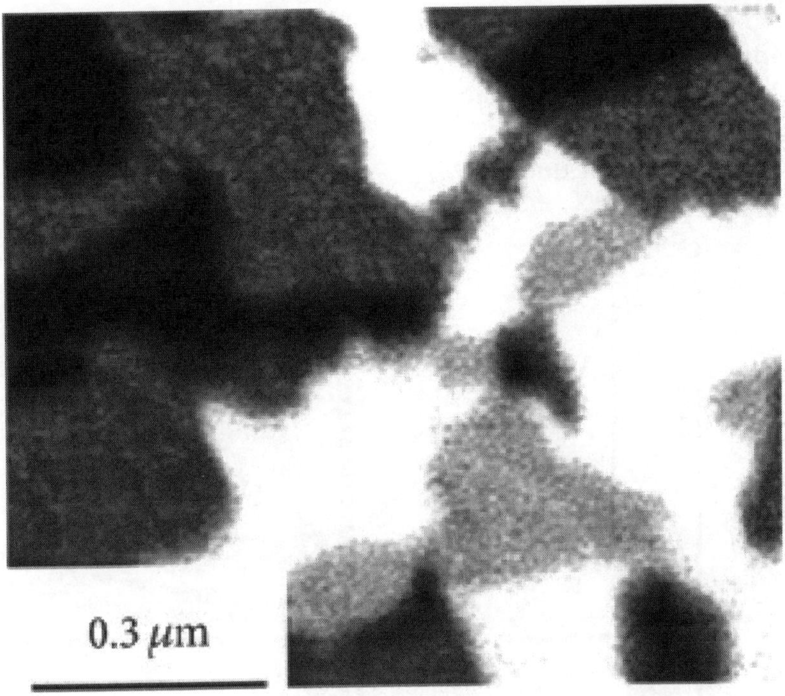

(f)

Figure 7: In-plane PFM images measured on (a) 120, (b) 15, and (c) 2 nm thick BFO samples on $SrTiO_3$ and in-plane PFM images measured on (d) 120, (e) 15, and (f) 2 nm thick BFO samples on $DySrO_3$ substrates.

PHYSICAL PROPERTIES OF LOW-DIMENSIONAL BFO NANOSTRUCTURES

BFO Nanoparticles

Park et al. [9] synthesized the BFO nanoparticles by sol-gel method and investigated the size-driven magnetism of the BFO nanoparticles, as shown in Figure 8. As the sizes of the BFO nanoparticles were reduced, their magnetism was significantly enhanced. Especially, when the particle size was 14 nm, the magnetization was about 3 times larger than that of the BFO particle with size of 100 nm. This is important for enhancing the magnetic properties of the BFO nanoparticles with antiferromagnetic ground state.

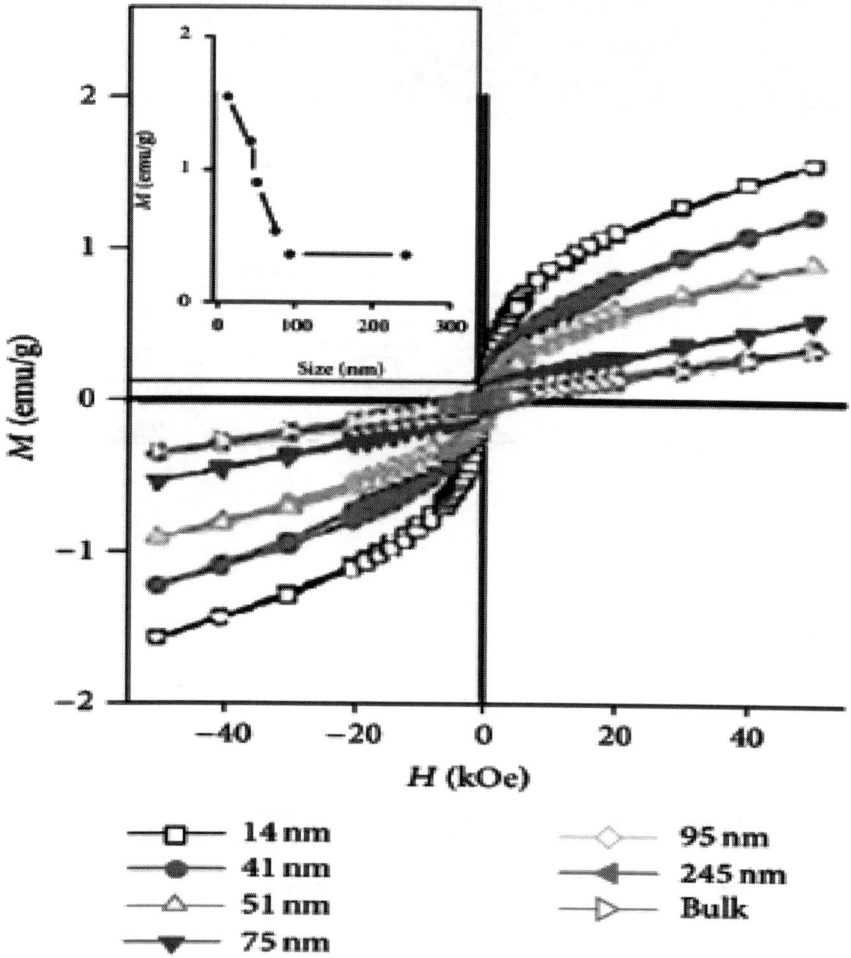

(a)

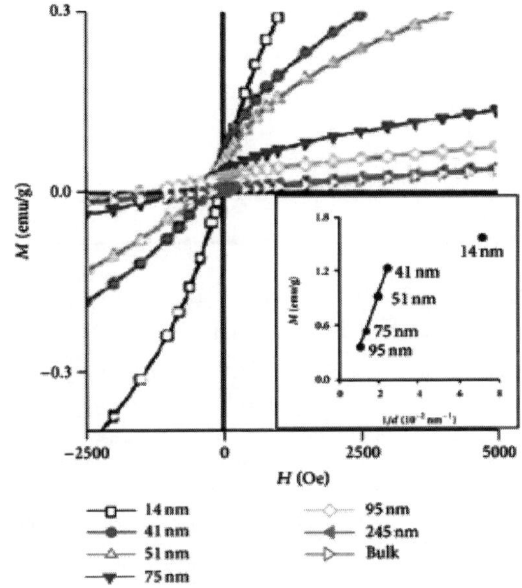

(b)

Figure 8: (a) Hysteresis loops measured at 300 K for $BiFeO_3$ nanoparticles with indicated sizes, (b) expanded plots of magnetization of $BiFeO_3$ nanoparticles with the return branches of the hysteresis loops omitted for clarity.

Recently it is reported that low-dimensional nanostructured BFO such as nanoparticles and nanowires exhibit good photocatalytic activities in visible-light region. As a novel visible-light-responsive photocatalysts for degradation of organic compounds BFO nanostructures have been widely investigated. For example, Zhu et al. [28] reported the microwave-hydrothermal synthesis of spherical perovskite-type BFO nanocrystals with diameters of 10–50 nm and hexagonal-shaped sillenite-type ones with sizes of 18–33 nm at low temperatures. They found that the sillenite-type bismuth ferritic nanocrystals exhibit higher photocatalytic ability than the perovskite-type ones, which was ascribed to their small mean particle sizes and the unique hexagonal-shape morphology, and also the structural characteristics of sillenite-type compound. Gao et al. [18] also reported visible-light photocatalytic property of BFO nanocrystals. As compared with the traditional TiO_2-based photocatalysts, which are the only response to UV irradiation due to its large band gap (3.2 eV), the BFO nanocrystals exhibit their obvious advantage making use of the visible-light due to their small band gaps. This is invaluable in increasing the photocatalytic reaction by using the visible sunlight.

BFO Nanoislands

Up to date, much work about the low-dimensional BFO nanostructures is mainly focused on the BFO nanoparticles (0D) and nanofilms (2D), and little work is reported on the BFO nanoislands. Geometrically, BFO nanoislands are a class of systems that bridge the gap between the BFO nanoparticles and BFO ultrathin films. Compared with BFO thin films, they have free-standing sidewalls that tend to suppress the formation of a nonuniform in-plane polarization due to the appearance of depolarizing field, similar to the ferromagnetic particles. On the other hand, relative to the (confined in all three dimensions) nanoparticles, the BFO nanoislands have large aspect ratio and likely to behave similarly to thin films when the polarization is out of plane. Therefore, it is expected that multiferroic BFO nanoislands should exhibit some kind of a "particle-to-thin film" crossover behavior and related novel effects depending on the aspect ratio and the type of bulk polarization ordering. Furthermore, due to the geometrical similarity between the multiferroic BFO nanoislands and microelectronic devices based on the multiferroic BFO nanomaterials, it is useful for simulating working conditions of real microelectronic devices [33, 34].

Recently, Hang et al. [29] reported the epitaxial growth of multiferroic BFO nanoislands on $SrTiO_3$ (100) and Nb-doped $SrTiO_3$ (100) single crystal substrates by chemical self-assembled method. By this method, they synthesized the epitaxial multiferroic BFO nanoislands via postannealing process in the temperature range of 650–800°C, and the lateral sizes of the BFO nanoislands were in the range of 50–160 nm and height of 6–12 nm. With increasing the postannealing temperature, the morphology of the BFO nanoislands in the (100) growth plane evolved from triangled to squared, and then to plated shapes. Ferroelectric characteristics of a single epitaxial BFO nanoisland (with lateral size of ~50 nm and height of 12 nm) grown on Nb-doped $SrTiO_3$(100) single crystal substrate was characterized by PFM images. The results demonstrated that fractal ferroelectric domains existed in the single BFO nanoisland, and self-biased polarization was also observed within this multiferroic nanoisland. This phenomenon can be ascribed to the interfacial stress caused by the lattice misfit between the BFO nanoisland and the $SrTiO_3$ single crystal substrate.

By using the top-down approach such as FIB milling method, Morelli et al. [27] fabricated the arrays of epitaxial BFO islands with flat top surfaces and lateral sizes down to 250 nm by starting from a continuous BFO thin film. PFM images showed that the as-fabricated BFO nanoislands preserved ferroelectric properties with switchable polarization and exhibited retention of polarization state at least for several days. As compared with the parent thin film, the BFO nanoislands exhibit a certain degree of imprint behavior, as shown in Figure9. That is due to the existence of the defects at the interface between the BFO film and $SrRuO_3$ substrate, and on the sidewalls of the islands.

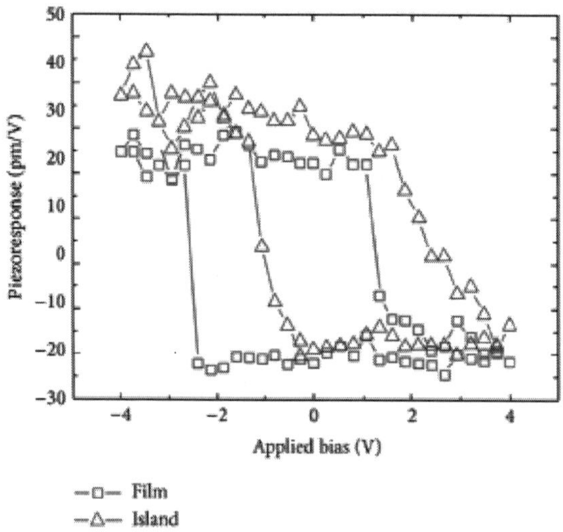

Figure 9: Local remanent piezoresponse hysteresis plots as resulting from measurements performed on an island (triangles) and on an area of the parent film (squares).

BFO Nanowires, Nanotubes, and Their Arrays

In the last decade low-dimensional BFO nanostructures have received much attention because of their superior physical and chemical properties. Among them, BFO nanowires and nanotubes are especially attractive for nanoscience studies and nanotechnology applications, which are ascribed to that the BFO nanowires and/or nanotubes are not only used as the building blocks of future nanodevices, but also offer fundamental scientific opportunities for investigating the intrinsic size effects of physical properties.

Nowadays BFO nanowires, nanorods, nanotubes, and their arrays have been fabricated by the template-aided synthesis. However, all the products prepared by this method exhibit polycrystalline structures due to the heterogeneous nucleation on the pore walls; there are very few reports on the synthesis of single crystalline nanowires through this method. To better understand the intrinsic size effects of physical properties, high quality of one-dimensional single crystalline BFO nanowires is highly required. Recently, Liu et al. [22] reported the synthesis of single-crystalline BFO nanowires (45–200 nm in diameter) by solvothermal method and measured their magnetic properties by superconducting quantum interference device (SQUID) at room temperature and low temperatures, as shown in Figure 10. Li et al. [35] also synthesized the BFO nanowires by solvothermal method (40–200 nm in diameter and several micrometers in length) and characterized a single BFO nanowire by PFM. The results shows the x and z-PFM hysteresis loops which clearly reveals the ferroelectric property of a single BFO nanowire. The BFO nanowire (with diameter of 20 nm) arrays are also fabricated by template-assisted sol-gel technique [36]. Their ferroelectric and dielectric properties are demonstrated in Figure 11. Figure 11(a) shows the polarization of BFO nanowire arrays as a function of applied electric field about 600 kV/cm with a frequency (v)=10 kHz at room temperature. The observed P-E hysteresis loop exhibits well-saturated rectangular shape due to the presence of less oxygen-related defects and phase purity of the nanowires. The high value of saturation polarization was around 54 μC/cm^2 observed at 535 kV/cm applied electric field. Figure 11(b) displays the relative dielectric constant ε of the BFO nanowire arrays as a function of frequency. A monotonous decrease of dielectric constant and dielectric loss was observed as Maxwell-Wagner type interfacial polarization, and in increasing the frequency, which was ascribed to good agreement with Koops phenomenological theory. The dielectric constant was measured to be as high as 492 at 1 KHz, which was due to space charge polarization resulting from the inhomogeneous dielectric structure. Recently, BFO nanotubes are also being prepared by a template synthesis involving the sol-gel technique [31]. Their ferroelectric and piezoelectric properties were characterized by PFM measurements. The piezoresponse d_{33} hysteresis loop of an individual BFO nanotube was measured using the conductive atomic force microscope tip applied with a 16.5 kHz ac electric field plus a swept dc voltage, and the result is shown in Figure 12(a). The decrease in d_{33} at high electric field, as shown in Figure 12(a), is ascribed to a consequence of the field-induced lattice hardening, which is typical for perovskite piezoelectrics. The significant piezoelectric characteristics illustrate the ferroelectric behavior of the BFO nanotubes. The dielectric constant and dielectric loss of the BFO nanotube arrays were also measured at room temperature as a function of the frequency in the range of 10^3–10^6 Hz, as shown in Figure 12(b). Both the dielectric constant and the dielectric loss show a remarkable decrease of up to 10^3 Hz and remain fairly constant afterward. The decrease in the dielectric constant with increasing the frequency represents the anomalous dispersion of the dielectric constant at low and intermediate frequencies, which has been explained by the phenomenon of dipole relaxation; while the variation in dielectric loss with frequency represents the relaxation absorption of the dielectrics.

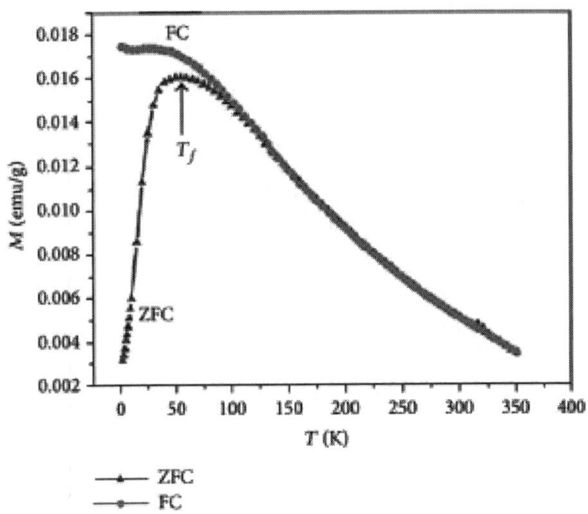

(a)

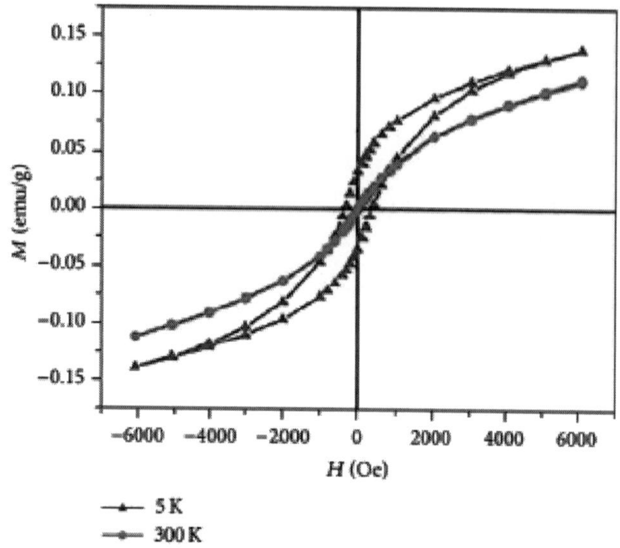

(b)

Figure 10: (a) Temperature dependence of zero-field-cooled (ZFC) and field-cooled (FC) susceptibility measured in a field of 100 Oe for $BiFeO_3$ nanowires, (b) M-H hysteresis loops for $BiFeO_3$ nanowires measured at 5 K and 300 K.

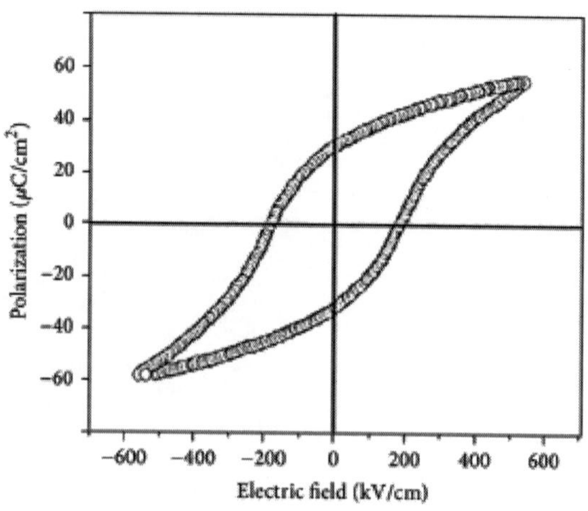

(a)

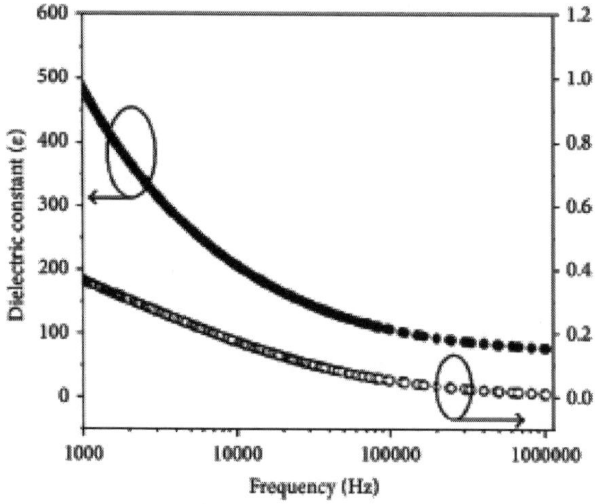

(b)

Figure 11: (a) Polarization versus applied electric field hysteresis loop, (b) relative dielectric constant and dielectric loss versus the frequency traits of $BiFeO_3$ nanowires.

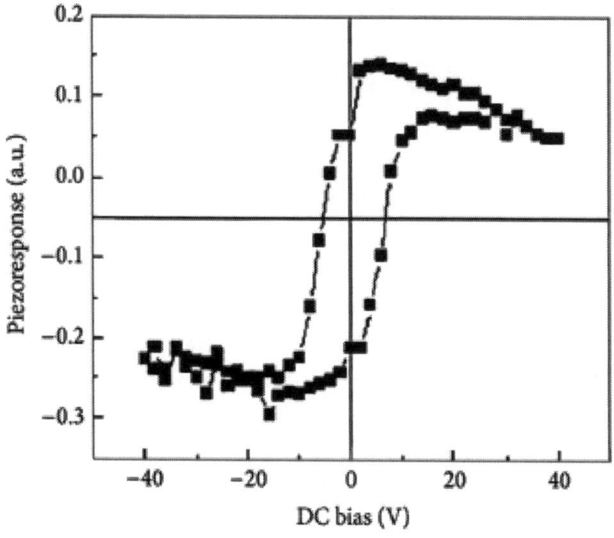

(a)

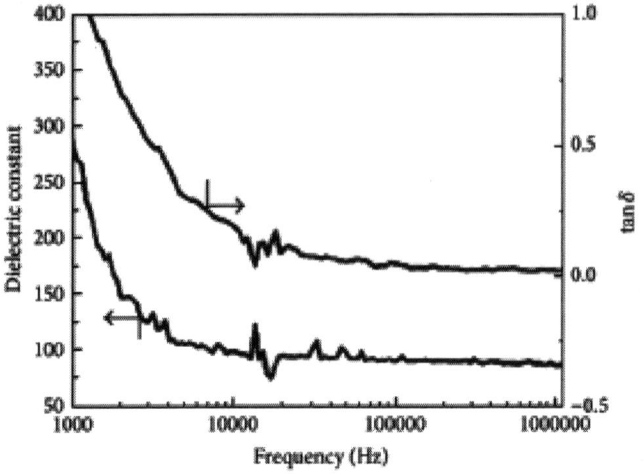

(b)

Figure 12: (a) Piezoelectric hysteresis loop of a single $BiFeO_3$ nanotube measured by PFM, (b) dielectric constant and dielectric loss measured at room temperature as a function of the frequency.

APPLICATIONS OF BFO NANOSTRUCTURES

BFO is one of several rare single-phase multiferroic materials that are both ferroelectric and weakly ferromagnetic at room temperature. Recent studies demonstrate that the BFO nanomaterials have spontaneous polarization enhancement, switchable ferroelectric diode effects, photovoltaic effects, piezoelectric, and THz radiation properties, which have potential applications in the fields of next-generation, lead-free, nondestructive memories, spin valve devices, actuators, and ultrahigh speed telecommunication devices [37].

BFO nanoparticles also exhibit good photocatalytic activities in visible-light region, which can be used as novel visible-light-responsive photocatalysts for degradation of organic compounds. For example, Zhu et al. [28] synthesized spherical perovskite-type single-crystalline BFO nanoparticles with diameters of 10–50 nm by microwave hydrothermal process, which exhibited efficient photocatalytic activity for the degradation of rhodamine B in aqueous solution under visible-light irradiation. Gao et al. [18] also synthesized BFO nanoparticles, which promoted the degradation rate of methyl orange to a high level under visible-light irradiation. In addition, Yu et al. [38] reported that perovskite-structured $BiFeO_3$ nanoparticles also exhibited excellent gas-sensing properties, which were potentially useful for high-quality gas sensors.

Due to the coupling of ferroelectric and antiferromagnetic vectors, reversing the ferroelectric polarization by an extern electric field also rotates the antiferromagnetic spins. Chu et al. [39] presented electric-field control of local ferromagnetism through the coupling between the multiferroic BFO and a ferromagnet (CoFe in their work). They grew heterostructures of Au (2 nm)/CoFe (2.5–20 nm)/BFO (50–200 nm)/ $SrRuO_3$ (25–50 nm) on $SrTiO_3$ (001)-oriented substrates. When applied an external in-plain electric field, they observed that the average magnetization direction in CoFe the ferromagnet rotates by 90°. The average magnetization direction changes back to the original state when applying the electric field again. If one suggests the original state as the binary signal "0," the rotated state of the magnetization direction is represented the signal "1." Therefore, the repeated such heterostructures can be used as essential building blocks to fabricate magnetoelectric random access memory elements.

Due to worldwide energy crisis, the investigation of materials for thin film photovoltaic cells is essential to renewable energy production. The large saturation polarization (~90 μC/cm^2) in BFO thin film and the band gap of BFO (E_g ~2.67eV) smaller than many other ferroelectric perovskites make BFO the remarkable candidate for the photovoltaic cells. Yang et al. [40] reported photovoltaic devices based on BFO thin films and demonstrated the highest efficiency for the ferroelectric-based photovoltaic. They grew epitaxial ferroelectric BFO thin film by metal-organic chemical vapor deposition on (001)-oriented $SrTiO_3$ substrates with 50 nm epitaxial $SrRuO_3$ as bottom electrodes. Figure 13(a) shows a set of polarization-electric field hysteresis loops as a function of the test frequency which reveals a strong diode-like behavior, characterized by a large, directional leakage at negative voltages. Figure 13(b) shows the I-V curves taken both in dark and under 285 mW/cm^{2}white-light illumination which reveals diode-like behavior and a photovoltaic effect of the heterostructure. External quantum efficiency

(EQE) measurement is showed in Figure 13(c) which reveals that the maximum conversion efficiency (~10%) is observed when the photon energy is larger than the band gap of BFO and drop-off at the shortest wavelength (<325 nm).

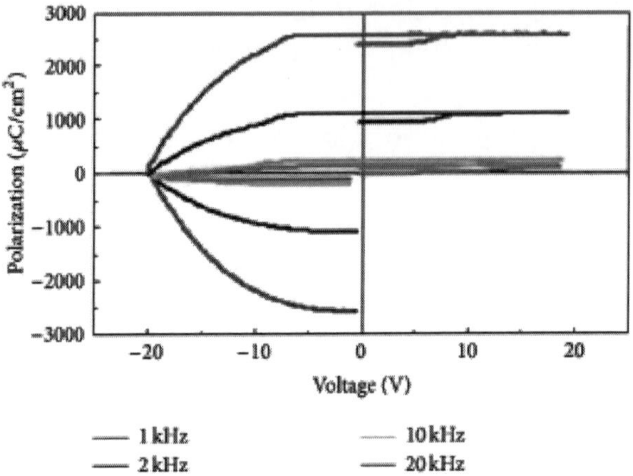

(a)

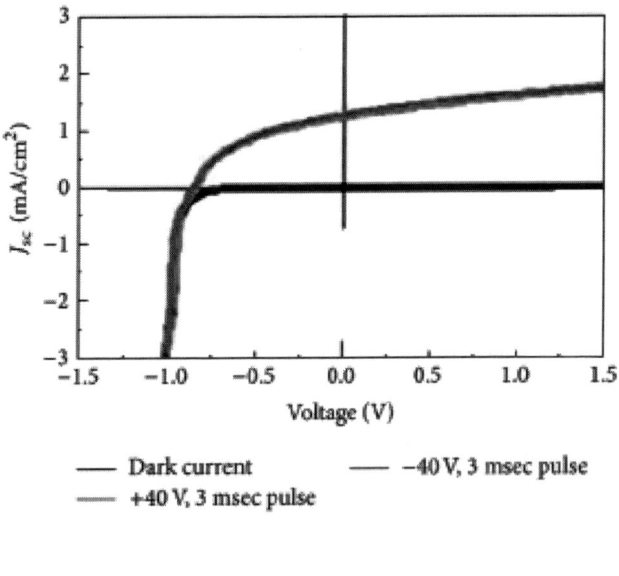

(b)

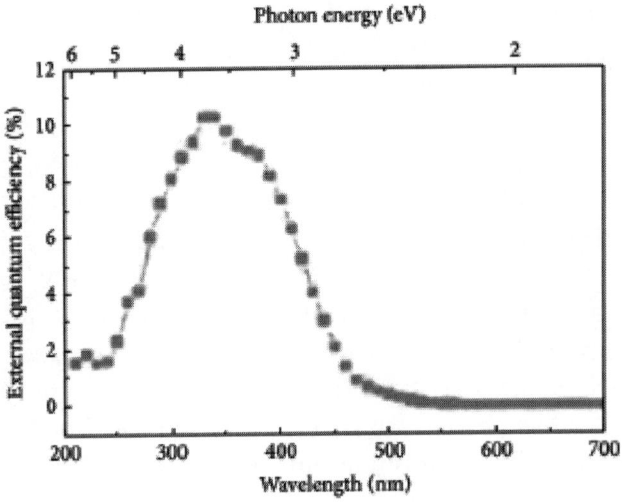

(c)

Figure 13: (a) Polarization-electric field hysteresis loops at various frequencies reveal diode-like behavior in one direction, (b) light (red and blue) and dark (black curve, running through the origin) I-V measurements completed at 2.85 suns intensity reveal photovoltaic effects in these device structures. There is no observed change in the light I-Vresponse upon application of an electric field and (c) average EQE measurements for five different contacts on a single sample reveal efficiencies ~10% under illumination with band gap light.

CONCLUSIONS

This paper reviews the recent research progress of low-dimensional BFO nanostructures, including their fabrication, property, structural characterization, and applications. Perovskite-type BFO as one of the few known single-phase multiferroics that possesses ferroelectricity and antiferromagnetism at room temperature. Its low-dimensional nanostructures are much attractive in the applications of multistate storage, magnetoelectric sensor, and spintronic devices. As the feature sizes of the microelectronic enter into nanoscale, there are still some problems that need to be solved in fabrication, characterization, and application of BFO low-dimensional nanostructures. For instance, in BFO nanostructures there exist quantum size effect, (ferroelectric and magnetoelectric) size effect, and surface/interface effect; all these effects must be considered together from experimental and theoretical researches, which are fundamental to developing the new generation of revolutionary electronic nanodevices. Although the BFO has good ferroelectricity, its weak ferromagnetism is highly required to be enhanced, which could be achieved in low-dimensional nanostructures. Therefore, deeper understanding of the fundamentals of the BFO low-dimensional nanostructures with the development of advanced technology and exploring the coexistrnce of ferroelectricity and ferromagnetism with strong coupling between them will be the future direction of BFO nanomaterials researches.

ACKNOWLEDGMENTS

This work was partially supported by National Natural Science Foundation of China (Grant nos. 11174122 and 11134004), National Basic Research Program of China (Grant nos. 2009CB929503 and 2012CB619400), and the open project from National Laboratory of Solid State Microstructures, Nanjing University.

REFERENCES

1. H. Schmid, "Multi-ferroic magnetoelectrics," Ferroelectrics, vol. 162, no. 1, pp. 317–338, 1994.
2. B. B. van Aken, J. R. Rivera, H. Schmid, and M. Fiebig, "Observation of ferrotoroidic domains," Nature, vol. 449, no. 7163, pp. 702–705, 2007.
3. N. A. Spaldin and M. Fiebig, "The renaissance of magnetoelectric multiferroics," Science, vol. 309, no. 5733, pp. 391–392, 2005.
4. C. Ederer and N. A. Spaldin, "Weak ferromagnetism and magnetoelectric coupling in bismuth ferrite,"Physical Review B: Condensed Matter and Materials Physics, vol. 71, no. 6, Article ID 060401, 4 pages, 2005.
5. C. Ederer and N. A. Spaldin, "Influence of strain and oxygen vacancies on the magnetoelectric properties of multiferroic

bismuth ferrite," Physical Review B, vol. 71, no. 22, Article ID 224103, 9 pages, 2005.

6. M. Kumar and K. L. Yadav, "Study of room temperature magnetoelectric coupling in Ti substituted bismuth ferrite system," Journal of Applied Physics, vol. 100, no. 7, Article ID 074111, 4 pages, 2006.
7. D. H. Wang, W. C. Goh, M. Ning, and C. K. Ong, "Effect of Ba doping on magnetic, ferroelectric, and magnetoelectric properties in mutiferroic $BiFeO_3$ at room temperature," Applied Physics Letters, vol. 88, no. 21, Article ID 212907, 3 pages, 2006.
8. T. Zhao, A. Scholl, F. Zavaliche et al., "Electrical control of antiferromagnetic domains in multiferroic $BiFeO_3$ films at room temperature," Nature Materials, vol. 5, no. 10, pp. 823–829, 2006.
9. T. Park, G. C. Papaefthymiou, A. J. Viescas, A. R. Moodenbaugh, and S. S. Wong, "Size-dependent magnetic properties of single-crystalline multiferroic $BiFeO_3$ nanoparticles," Nano Letters, vol. 7, no. 3, pp. 766–772, 2007
10. H. Zhang and K. Kajiyoshi, "Hydrothermal synthesis and size-dependent properties of multiferroic bismuth ferrite crystallites," Journal of the American Ceramic Society, vol. 93, no. 11, pp. 3842–3849, 2010.
11. R. R. Das, D. M. Kim, S. H. Baek et al., "Synthesis and ferroelectric properties of epitaxial $BiFeO_3$ thin films grown by sputtering," Applied Physics Letters, vol. 88, no. 24, Article ID 242904, 3 pages, 2006.
12. R. Mazumder, P. S. Devi, D. Bhattacharya, P. Choudhury, A. Sen, and M. Raja, "Ferromagnetism in nanoscale $BiFeO_3$," Applied Physics Letters, vol. 91, no. 6, Article ID 062510, 3 pages, 2007.
13. S. M. Stratulat, X. L. Lu, A. Morelli, D. Hesse, W. Erfurth, and M. Alexe, "Nucleation-induced self-assembly of multiferroic $BiFeO_3$- $CoFe_2O_4$ nanocomposites," Nano Letters, vol. 13, no. 8, pp. 3884–3889, 2013.
14. R. Nath, S. Hong, J. A. Klug et al., "Effects of cantilever buckling on vector piezoresponse force microscopy imaging of ferroelectric domains in $BiFeO_3$ nanostructures," Applied Physics Letters, vol. 96, no. 16, Article ID 163101, 3 pages, 2010.

15. M. Park, S. Hong, J. A. Klug et al., "Three-dimensional ferroelectric domain imaging of epitaxial $BiFeO_3$ thin films using angle-resolved piezoresponse force microscopy," Applied Physics Letters, vol. 97, no. 11, Article ID 112907, 3 pages, 2010.
16. R. Nath, S. Hong, J. A. Klug et al., "Effects of cantilever buckling on vector piezoresponse force microscopy imaging of ferroelectric domains in $BiFeO_3$ nanostructures," Applied Physics Letters, vol. 96, Article ID 163101, 2010.
17. J. K. Kim, S. S. Kim, and W. J. Kim, "Sol-gel synthesis and properties of multiferroic $BiFeO_3$," Materials Letters, vol. 59, no. 29-30, pp. 4006–4009, 2005.
18. F. Gao, X. Chen, K. Yin et al., "Visible-light photocatalytic properties of weak magnetic $BiFeO_3$ nanoparticles," Advanced Materials, vol. 19, no. 19, pp. 2889–2892, 2007.
19. X.-Z. Chen, Z.-C. Qiu, J.-P. Zhou, G. Zhu, X.-B. Bian, and P. Liu, "Large-scale growth and shape evolution of bismuth ferrite particles with a hydrothermal method," Materials Chemistry and Physics, vol. 126, no. 3, pp. 560–567, 2011.
20. S. H. Han, K. S. Kim, H. G. Kim et al., "Synthesis and characterization of multiferroic $BiFeO_3$ powders fabricated by hydrothermal method," Ceramics International, vol. 36, no. 4, pp. 1365–1372, 2010.
21. Y. Wang, G. Xu, Z. Ren et al., "Mineralizer-assisted hydrothermal synthesis and characterization of $BiFeO_3$ nanoparticles," Journal of the American Ceramic Society, vol. 90, no. 8, pp. 2615–2617, 2007.
22. B. Liu, B. Hu, and Z. Du, "Hydrothermal synthesis and magnetic properties of single-crystalline $BiFeO_3$ nanowires," Chemical Communications, vol. 47, no. 28, pp. 8166–8168, 2011.
23. L. Zhang, X. Cao, Y. Ma, X. Chen, and Z. Xue, "Polymer-directed synthesis and magnetic property of nanoparticles-assembled $BiFeO_3$ microrods," Journal of Solid State Chemistry, vol. 183, no. 8, pp. 1761–1766, 2010.
24. T. J. Park, Y. Mao, and S. S. Wong, "Synthesis and characterization of multiferroic $BiFeO_3$ nanotubes,"Chemical Communications, no. 23, pp. 2708–2709, 2004.

25. X. Y. Zhang, J. Y. Dai, and C. W. Lai, "Synthesis and characterization of highly ordered $BiFeO_3$multiferroic nanowire arrays," Progress in Solid State Chemistry, vol. 33, no. 2–4, pp. 147–151, 2005.

26. F. Gao, Y. Yuan, K. F. Wang et al., "Preparation and photoabsorption characterization $BiFeO_3$nanowires," Applied Physics Letters, vol. 89, no. 10, Article ID 102506, 3 pages, 2006.

27. A. Morelli, F. Johann, N. Schammelt, D. McGrouther, and I. Vrejoiu, "Mask assisted fabrication of nanoislands of $BiFeO_3$ by ion beam milling," Journal of Applied Physics, vol. 113, no. 15, Article ID 154101, 4 pages, 2013.

28. X. H. Zhu, Q. M. Hang, Z. B. Xing et al., "Microwave hydrothermal synthesis, structural characterization, and visible-light photocatalytic activities of single-crystalline bismuth ferric nanocrystals," Journal of the American Ceramic Society, vol. 94, no. 8, pp. 2688–2693, 2011.

29. Q. M. Hang, X. H. Zhu, Z. J. Tang, Y. Song, and Z. G. Liu, "Self-assembled perovskite epitaxial multiferroic $BiFeO_3$ nanoislands," Advanced Materials Research, vol. 197-198, pp. 1325–1331, 2011.

30. J. Zhou, S. Liang, S. Y. Li, Z. D. Liu, Y. Y. Zhu, and X. H. Zhu, "Advances in low-dimensional $BiFeO_3$multiferroic nanostructures," Journal of Chinese Electron Microscopy Society, vol. 32, no. 6, pp. 504–524, 2013.

31. X. Y. Zhang, C. W. Lai, X. Zhao, D. Y. Wang, and J. Y. Dai, "Synthesis and ferroelectric properties of multiferroic $BiFeO_3$ nanotube arrays," Applied Physics Letters, vol. 87, no. 14, Article ID 143102, 3 pages, 2005.

32. Y. H. Chu, T. Zhao, M. P. Cruz et al., "Ferroelectric size effects in multiferroic $BiFeO_3$ thin films,"Applied Physics Letters, vol. 90, no. 25, Article ID 252906, 3 pages, 2007.

33. I. Naumov and A. M. Bratkovsky, "Unusual polarization patterns in flat epitaxial ferroelectric nanoparticles," Physical Review Letters, vol. 101, no. 10, Article ID 107601, 4 pages, 2008.

34. S. K. Streiffer and D. D. Fong, "Phase transitions in nanoscale ferroelectric structures," MRS Bulletin, vol. 34, no. 11, pp. 832–837, 2009.

35. S. Li, R. Nechache, C. Harnagea, L. Nikolova, and F. Rosei, "Single-crystalline $BiFeO_3$ nanowires and their ferroelectric behavior," Applied Physics Letters, vol. 101, no. 19, Article ID 192903, 3 pages, 2012.

36. G. S. Lotey and N. K. Verma, "Magnetoelectric coupling in multiferroic $BiFeO_3$ nanowires," Chemical Physics Letters, vol. 579, no. 30, pp. 78–84, 2013.

37. M. Bibes and A. Barthélémy, "Multiferroics: towards a magnetoelectric memory," Nature Materials, vol. 7, no. 6, pp. 425–426, 2008.

38. X. L. Yu, Y. Wang, Y. M. Hu, C. Cao, and H. L. Chan, "Gas-sensing properties of perovskite $BiFeO_3$nanoparticles," Journal of the American Ceramic Society, vol. 92, no. 12, pp. 3105–3107, 2009.

39. Y. H. Chu, L. W. Martin, M. B. Holcomb et al., "Electric-field control of local ferromagnetism using a magnetoelectric multiferroic," Nature Materials, vol. 7, no. 6, pp. 478–482, 2008.

40. S. Y. Yang, L. W. Martin, S. J. Byrnes et al., "Photovoltaic effects in $BiFeO_3$," Applied Physics Letters, vol. 95, no. 6, Article ID 062909, 3 pages, 2009.

Chapter 5

Research Progress on Electronic Phase Separation in Low-Dimensional Perovskite Manganite Nanostructures

Lizhi Liang, Lei Li, Heng Wu, and Xinhua Zhu

National Laboratory of Solid State Microstructures, School of Physics, Nanjing University, Nanjing 210093, China

ABSTRACT

Perovskite oxide manganites with a general formula of $R_{1-}xAxMnO_3$ (where R is a trivalent rare-earth element such as La, Pr, Sm, and A is a divalent alkaline-earth element such as Ca, Sr, and Ba) have received much attention due to their unusual electron-transport and magnetic properties, which are indispensable for applications in microelectronic, magnetic, and spintronic devices. Recent advances in the science and technology have resulted in the feature sizes of microelectronic devices based on perovskite manganite oxides down-scaling into nanoscale dimensions. At the nanoscale, low-dimensional perovskite manganite oxide nanostructures display novel physical properties that are different from their bulk and film counterparts. Recently, there is strong experimental evidence to indicate that the low-dimensional perovskite manganite oxide nanostructures are electronically inhomogeneous, consisting of different spatial regions with different electronic orders, a phenomenon that is named as electronic phase separation (EPS). As the geometry sizes of the low-dimensional manganite nanostructures are reduced to the characteristic EPS length scale (typically several tens of nanometers in manganites), the EPS is expected to be strongly modulated, leading to quite dramatic changes in functionality and more emergent phenomena. Therefore, reduced dimensionality opens a door to the new functionalities in perovskite manganite oxides and offers a way to gain new insight into the nature of EPS. During the past few years, much progress has been made in understanding the physical nature of the EPS in low-dimensional perovskite manganite nanostructures both from experimentalists and theorists, which have a profound impact on the oxide nanoelectronics. This nanoreview covers the research progresses of the EPS in low-dimensional perovskite manganite nanostructures such as nanoparticles, nanowires/nanotubes, and nanostructured films and/or patterns. The possible physical origins of the EPS are also discussed from the signatures of electronic inhomogeneities as well as some theoretical scenarios, to shed light on understanding this phenomenon. Finally, the perspectives to the future researches in this area are also outlined.

REVIEW

Background

Strongly correlated-electron materials, such as the rare-earth perovskite oxide manganites having a general formula $R_{1-x}AxMnO_3$, where R is a trivalent rare-earth element (e.g., La, Pr, Sm) and A is a divalent alkaline-earth element such as Ca, Sr, and Ba, have been attracting much attention because of their unusual electron-transport and magnetic properties, e.g., colossal magnetoresistance (CMR) effect [1-3], a sharp metal-insulator transition (MIT) as a function of temperature, electric field, magnetic field, light, hydrostatic pressure, strain, etc. [4-6]. Such MIT is also accompanied by a paramagnetic to ferromagnetic transition as the temperature is lowered. The competition between several interactions in the rare-earth perovskite oxide manganites makes that only small energy differences exist between the different possible phases of the system. As a result, the phase of the material can be tuned by various external perturbations, such as magnetic and electric fields, strain, and disorder. These perturbations may lead to the CMR effect and can be used for electronic phase control in manganite devices. Recently, there is strong experimental evidence to indicate that the rare-earth perovskite oxide manganites are electronically inhomogeneous, which consist of different spatial regions with different electronic orders [7-10], a phenomenon that is named as electronic phase separation (EPS). As an inherent electronic inhomogeneity, EPS has been widely reported in the rare-earth perovskite oxide manganites, and its size varies from nano to mesoscopic scales [11-15]. It has been recognized to be crucial for the CMR effect and the MIT in manganites, leading to the new applications of spintronics [9]. However, the presence of EPS raises many intriguing questions, e.g., what is the microscopic nature of the EPS? Why does it have such a large range of length scales from nanometers to micrometers? More importantly, is it responsible for the related physical properties such as CMR and high-Tc superconducting exhibited by the manganites and related oxide materials? Therefore, EPS is getting recognized as a phenomenon of importance in understanding the magnetic and electron transport properties of perovskite oxide manganites[16,17].

Recent advances in science and technology of perovskite oxide manganites have resulted in the feature sizes of the microelectronic devices based on perovskite oxide manganites entering into nanoscale dimensions. At nanoscale, perovskite oxide manganites exhibit a pronounced size effect manifesting itself in a significant deviation of the properties of low-dimensional structures from their bulk and film counterparts. For example, in the charge-ordered half-doped manganite nanoparticles, a robust charge ordering (CO) state and antiferromagnetic (AFM) ground state are significantly suppressed whereas a ferromagnetic (FM) ordering is exhibited [18-22]; meanwhile, there exists size-dependent exchange bias effect and glassy behaviors [23-26]. In the nanostructured patterns of (La,Pr,Ca) MnO_3 (LPCMO) narrow strips (spatial confined system), several new transport features such as giant resistance jumps [27-30], reentrant M-I transitions [31], negative differential resistances, and intrinsic tunneling magnetoresistance [32,33] emerge, which are absent in the thin films and bulks. Furthermore, as the geometry size of the low-dimensional manganite nanostructures is further reduced to the characteristic EPS length scale (typically several tens of nanometers in manganites), the EPS is expected to be strongly modulated, leading to quite dramatic changes in functionality and more emergent phenomena [34]. Therefore, reduced dimensionality will open a door to the new functionalities in perovskite manganites and offer a way to gain new insight into the nature of EPS in the perovskite manganite system [35]. In the recent years, much progress has been made in understanding the physical nature of the EPS in low-dimensional perovskite manganite nanostructures both from experimentalists and theorists, which have a profound impact on the manganite oxide nanoelectronics. In this work, we review the major progress of the EPS in low-dimensional perovskite manganite nanostructures, which are based on the recent literatures about the EPS in perovskite manganite nanoparticles, nanowires/ nanotubes, and nanostructured films and/or patterns. The possible physical origins of the EPS are also discussed from the signatures of electronic inhomogeneities as well as some theoretical scenarios to shed light on understanding this phenomenon. We end this review by providing our perspectives to the future research directions in this area.

Research History of EPS and its Signatures

The first report on the EPS in perovskite manganites was back to 1950s, where Wollan and Koehler carried out their pioneering neutron scattering studies of $La_{1-x}CaxMnO_3$ (LCMO) [36]. They observed both FM and AFM peaks in the magnetic structure of LCMO by neutron scattering, and concluded that there is the simultaneous presence of FM and AFM phases in this material. Since that time, manganites had just begun to attract the interest of physicists. In 1950, Jonker and van Santen first reported the electrical and magnetic properties of manganites, and they found a ferromagnetic conducting phase below room temperature in $La_{1-x}CaxMnO_3$ $(0.2<x<0.4)$ [37,38]. Shortly afterward, Zener, Kanamori, Goodenough, and several others established the basic theoretical framework of the EPS that scientists use today [39].

Manganites and the phase separation effects they display fell out of fashion until the 1990s. Although significant magnetoresistance effects in single-crystal $La_{0.69}Pb_{0.31}MnO_3$ were reported in 1969, there was no technological incentive for further pursuit [40]. What reignited manganite research was the 1994 discovery by Jin and collaborators that a several-tesla magnetic field could induce a 1,000-fold change in the resistance of a heat-treated epitaxial thin film of $La_{0.67}Ca_{0.33}MnO_3$ [41]. A dubbed CMR, this effect arises because the applied magnetic field drives a phase transition from an insulating paramagnet to a spin-aligned metal. Thus, as Jonker and van Santen reduced the temperature to reach the conducting spin-aligned phase, Jin and his colleagues applied a magnetic field.

Recently, Woodward et al. performed a neutron diffraction study of $Nd_{0.5}Sr_{0.5}MnO_3$ and found that this material first became FM at 250 K, partially transforming to an A-type AFM phase at approximately 220 K, followed by a transformation of a substantial fraction to a CE-type AFM phase at approximately 150 K [42]. Their experimental results indicate that three phases (FM metallic and CE-AFM charge-ordered phases along with an A-type AFM phase) coexist at low temperatures, and the size scale of the inhomogeneities is at least in the mesoscopic range (a few hundred nanometres or more). Sub-micrometersized phase separation involving FM and charge-ordered AFM domains with a typical size of about 0.2 μm was found in $La_{0.625-y}Pr_yCa_{0.375}MnO_3$ by transmission electron microscopy (TEM) [5]. At the same time, by using scanning tunneling spectroscopy (STM), Fäth et al. also found the evidence of electronic inhomogeneities in $La_{0.7}Ca_{0.3}MnO_3$ below the FM transition temperature with a mesoscopic scale of about 0.2 μm, where the FM metallic domains are interspersed in insulating regions [43].

Mesoscopic phase separation with the length scale between 30 and 200 nm, arising from the comparable energies of the ferromagnetic metallic and antiferromagnetic insulating states, is just one extreme in the perovskite manganites [5]. Normally, the EPS with phases of different charge densities is expected to give rise to nanometer scale clusters because large phase separated domains would break up into small pieces due to the Coulomb interactions. For example, Mori et al.

reported a nanoscopic length scale of the electronic inhomogeneity in thin films of the hole-doped side of (La,Ca)MnO_3 by high-resolution TEM [44]. Similarly, in $Bi_{0.25}Ca_{0.75}MnO_3$, Renner et al. also found nanoscopic charge-ordered and metallic domains which were correlated with the structural distortions[45]. Generally, microscopically homogeneous clusters are usually in the diameter size of 1 to 2 nm dispersed in an insulating or charge-localized matrix. For example, recently, De Teresa et al. [46] reported on the experimental evidence for the existence of nanoscopic phase segregation in the manganite compounds of $(La_{1-x}A_x)_{2/3}Ca_{1/3}MnO_3$ (A=Y or Tb), in which the spontaneous formation of localized magnetic clusters with size of ~1.2 nm above the ferromagnetic ordering temperature was revealed by a combination of volume thermal expansion, magnetic susceptibility, and small-angle neutron scattering measurements. These nanosized magnetic clusters grew in size but decreased in numbers under an external applied magnetic field. Such a phase separation scenario bridges the gap between the double-exchange model and the lattice distortion models.

The signatures of EPS can be revealed by different techniques depending on the length scale on which it occurs. For mesoscopic phase separation, diffraction techniques can be used to reveal its distinct features since the size scale of the inhomogeneities is large enough to produce well-defined reflections in neutron and X-ray diffraction patterns [9]. However, for the nanoscopic electronic inhomogeneity in manganites, both TEM, high-resolution TEM and scanning transmission electron microscopy (STEM), and STM can be used to reveal the coexistence of nanoscopic charge-ordered (insulating) domains and the FM metallic domains, giving the local structural information at atomic level [5]. It is often difficult to identify EPS based on the magnetization and transport measurements because of the sensitivity of phase separation to magnetic fields. Thus, magnetic fields transform the antiferromagnetic insulating state to the ferromagnetic metallic state. However, transport measurements, under favorable conditions, can provide valuable information on phase separation.

EPS in Low-dimensional Perovskite Manganite Nanostructures

Over the last decade, nanomaterials have received much attention from the scientific and engineering viewpoints. They exhibit different properties from those of bulk materials due to their small size and large surface-to-volume ratios, and become promising candidates for nanometer scale electronic, optical, and mechanical devices. Recent advances in science and technology of perovskite manganites have resulted in the feature sizes of perovskite manganite-based oxide electronic devices entering into nanoscale dimensions. As the spatial dimension of the low-dimensional manganite nanostructures is reduced to the characteristic EPS length scale, quite dramatic changes in their transport properties such as ultrasharp jumps of magnetoresistance, reentrant MIT, negative differential resistances, and intrinsic tunneling magnetoresistance could appear, which are believed to be caused in large part by the EPS in perovskite manganite nanostructures [27-33]. They have significant impacts on fabricating oxide-based novel devices. To better understand the EPS phenomenon in low-dimensional perovskite manganite nanostructures, in the past several years, various synthetic methods such as sol–gel technique [47], hydrothermal synthesis [48,49], electro-spinning process [50,51], template method[52-54], and lithographic techniques [27,29-31,33,34] have been developed to fabricate low-dimensional manganite nanostructures, such as manganite nanoparticles, nanowires/nanotubes, and nanostructured films/patterns. In addition, the new technologies to directly identify the EPS phenomenon in low-dimensional perovskite manganite nanostructures have been developed, such as scanning electron nanodiffraction [13], atomic-resolution STM [45], MFM [4,55], and electron holography [11]. In this section, we will review the recent progress on the EPS in low-dimensional perovskite manganite nanostructures.

EPS in Manganite Nanoparticles

EPS is an important phenomenon in CMR material, which leads to the new applications of spintronics. Along with the development of nanotechnology, the EPS phenomenon in CMR nanoparticles are received much attention. Recently, the evolution of the EPS with magnetic field in nanosized $Nd_{0.5}Ca_{0.5}MnO_3$[19], $La_{0.25}Ca_{0.75}MnO_3$[47], $Pr_{0.5}Ca_{0.5}MnO_3$[21], $La_{0.2}Ca_{0.8}MnO_3$[56], and $Pr_{0.67}Ca_{0.33}MnO_3$[57] particles has been reported. For example, in nanosized $Pr_{0.67}Ca_{0.33}MnO_3$particles with average diameter of 100 nm, it was found that a sharp transition from AFM to FM did not occur even up to 60 kOe, as demonstrated in Figure 1 [57]. The field dependence of the analyzed magnetization data for the $Pr_{0.67}Ca_{0.33}MnO_3$ nanoparticles is shown in Figure 2[57]. As a comparison, the data for the bulk counterpart is also given out. It is clear that the evolution tend of M is a little different from that of the bulk counterpart, i.e., first a sharp decrease and then an increase slowly up to 50 kOe. However, the irreversibility temperature (T_{irr}) exhibits a very different change tend as compared with that of the bulk counterpart, which is sharply decreased from 100 to 5,000 Oe and then continually increased. The magnetization M_{ZFC} and M_{FC} are increased smoothly with increasing the magnetic field H up to 60 kOe but a step-like increase of M_{ZFC} and M_{FC} like in bulk counterpart is not observed. For H below 5,000 Oe, the first sharp decrease of the M T_{irr}, and weak decline of T is attributed to the gradual conquest of the anisotropy of frozen spin and alignment with field, since the magnetic field is not large enough to induce the growth of the FM cluster. Due to the surface effect, the FM-like surface spins contribute additional moment, which leads to a large magnetization for nanoparticles as compared with bulk counterpart. However, due to the strong coupling between the surface spins and interface spins (which also deviate from AFM arrangement), the exchange field required to force a transition of surface spins and interface spins to full FM is approximately 5×10^6 Oe [58]. As a consequence, even the field is increased up to 60 kOe, which can align the AFM core spins like for bulk, it is still not large enough to make the nanoparticles to be full FM configuration, thus leading to a slow increase of the M and T_{irr} [58]. The significant increase of the exchange bias field of the $Pr_{0.67}Ca_{0.33}MnO_3$ nanoparticle as compared with the bulk counterpart can be attributed to the surface pressure and uncompensated surface spins.

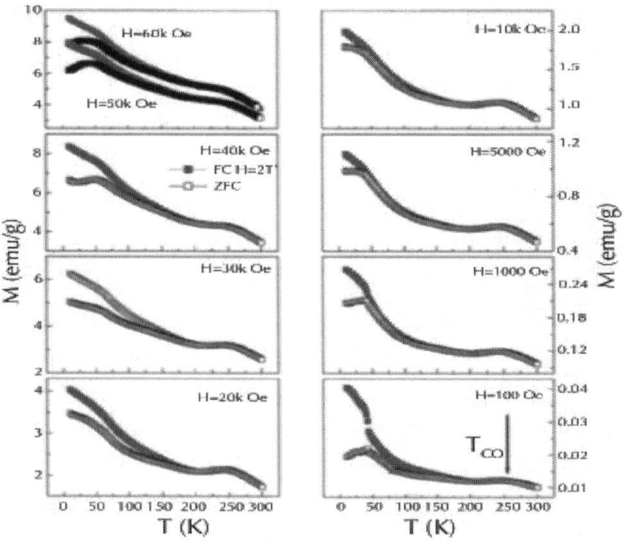

Figure 1: Field cooled and zero field cooled magnetization of Pr0.67Ca0.33MnO3 nanoparticles. Field cooled (closed symbols) and zero field cooled (open symbols) magnetization of $Pr_{0.67}Ca_{0.33}MnO_3$nanoparticles as a function of temperature, measured in the field ranges from 100 Oe to 60 kOe [57].

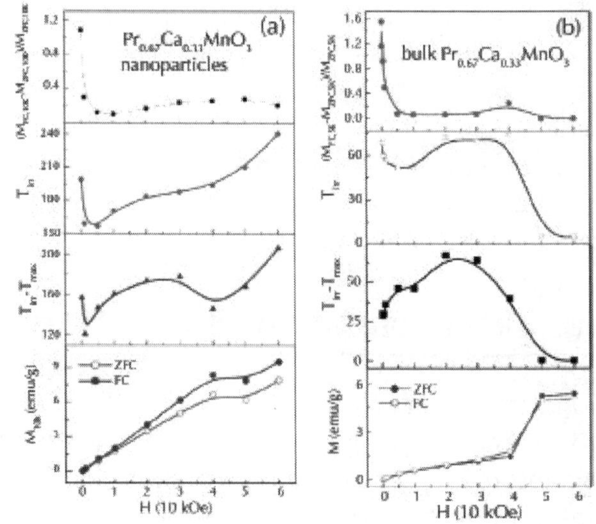

Figure 2: Field dependence of the analyzed magnetization data for (a) Pr0.67Ca0.33MnO3 nanoparticles and (b) bulk counterpart [[57]].The rela-

tive history dependence of the magnetization $M=(M_{FC}-M_{ZFC})/M_{ZFC}$ was measured at 10 K for $Pr_{0.67}Ca_{0.33}MnO_3$ nanoparticles and 5 K for bulk counterpart. T_{irr} is the irreversibility temperature; $T=T_{irr}-T_{max}$ is the difference between the irreversibility temperature and the temperature of the maximum ZFC magnetization. M_{ZFC} and M_{FC} at 10 K for $Pr_{0.67}Ca_{0.33}MnO_3$nanoparticles and 5 K for bulk counterpart.

Recently, the EPS in $La_{0.7}Sr_{0.3}MnO_3$ nanoparticles synthesized by sol–gel process was also investigated by electron magnetic resonance (EMR) method [59]. The results showed that all the $La_{0.7}Sr_{0.3}MnO_3$ nanoparticles (synthesized with different gelation agents) exhibited the following common features: (i) at the PM region, the EMR line was pure Lorentzian having a *g* value decreasing with increasing the temperature and *g* value reached 2 at around 350 K; (ii) when the temperatures are crossing Tc, the EMR lines changed their resonance fields (e.g., lineshapes and linewidths); (iii) all samples showed the coexistence of FM and PM signals within a wide temperature range below Tc; and the intensity of PM signal increased gradually as the temperature approached to Tc. The growth of PM phase was accompanied by a consequent decrease of FM signal intensity. Besides these common features, the EMR spectra of the measured samples also show several significant differences, which allow ones to investigate the origin of PS in these samples. It was found that the $La_{0.7}Sr_{0.3}MnO_3$ nanoparticles synthesized with different gelation agents in sol–gel process exhibited different magnetic behaviors, and a sharp FM-PM transition was observed in the $La_{0.7}Sr_{0.3}MnO_3$nanoparticles synthesized with a combined agent of urea and trisodium citrate. These results also demonstrate that the synthesis conditions of perovskite manganite nanoparticles have an important role in their microstructure, magnetic properties, and phase separation behavior.

EPS in Manganite Nanowires/Nanotubes

One-dimensional manganite nanostructures that include nanowires, nanorods, and nanotubes have attracted rapidly growing interest due to their fascinating electrical and magneto-transport properties. They are emerging as important building blocks serving as interconnects and active components in nanoscale electronic, magnetic, and spintronic devices. It is expected that the manganite nanowires will exhibit an emerging magnetic and transport behaviors associated the EPS due to the strong electronic correlation under a spatial confinement in the case of nanowires [35]. Recently, theoretical calculations using the FM Kondo Hamiltonian have predicted that the intrinsic EPS persists in one-dimensional manganite nanostructures [60]. However, due to the growth difficulty in single-crystalline manganite nanowires, the EPS in manganite nanowires are not completely explored. Recently, it has been demonstrated that by utilizing MgO nanowires as the template one can grow the transition metal oxide core-shell nanowires with good single crystalline quality [61,62]. By the same method, Li et al. synthesized the single-crystalline $La_{0.33}Pr_{0.34}Ca_{0.33}MnO_3$ (LPCMO)/MgO core-shell nanowires with diameters about tens of nanometers [63]. Their structure and morphology characterizations confirm the epitaxial growth of $La_{0.33}Pr_{0.34}Ca_{0.33}MnO_3$ shell layers on MgO core layers. The magnetic measurements are shown in Figure 3[63]. As shown in Figure 3a, the ZFC curve and the FC curve of the LPCMO nanowires are split at a blocking temperature of $T_b = 93$ K when the temperature is decreased. Such a ZFC/FC deviation is very similar to that of the bulk polycrystalline LPCMO sample also shown in Figure 3a, and is due to the frozen of the magnetic moment. The differences between the ZFC and FC magnetic moments in the nanowire, defined as the frozen phase magnetic moment, is significantly larger than that in the bulk counterpart below the blocking temperature sample, as shown in Figure 3b. In bulk or thin film LPCMO, the frozen phase is generally regarded to be related to the phase competition between the FM metallic phase and the AFM-CO phase [64]. So, in the nanowires, the increased amount of frozen phase concentration could be reasonable due to the stronger phase competition in the low-dimensional system. Figure 3c,d displays the magnetic field dependence of the magnetic moments of the LPCMO nanowires and the bulk counterpart. As observed in Figure 3c both the saturation magnetic moment m_s and the coercivity H_c in the LPCMO nanowires were increased as the temperature was decreased, which was similar to that in bulk or thin-film manganites. However, the differences between the nanowire and the bulk sample were also observed. The H_c value of the LPCMO nanowires was much larger than that of the LPCMO bulk sample. For example, at $T = 10$ K, H_c is about 550 Oe in the nanowire but only about 100 Oe in the bulk sample as shown in Figure 3d. The larger H_c in the nanowires could be attributed to their stronger domain wall pinning at the boundaries of the separated AFM and FM phases caused by the EPS in the nanowires

[65]. These observations suggest that the EPS with a stronger phase competition exists in the one-dimensional structure.

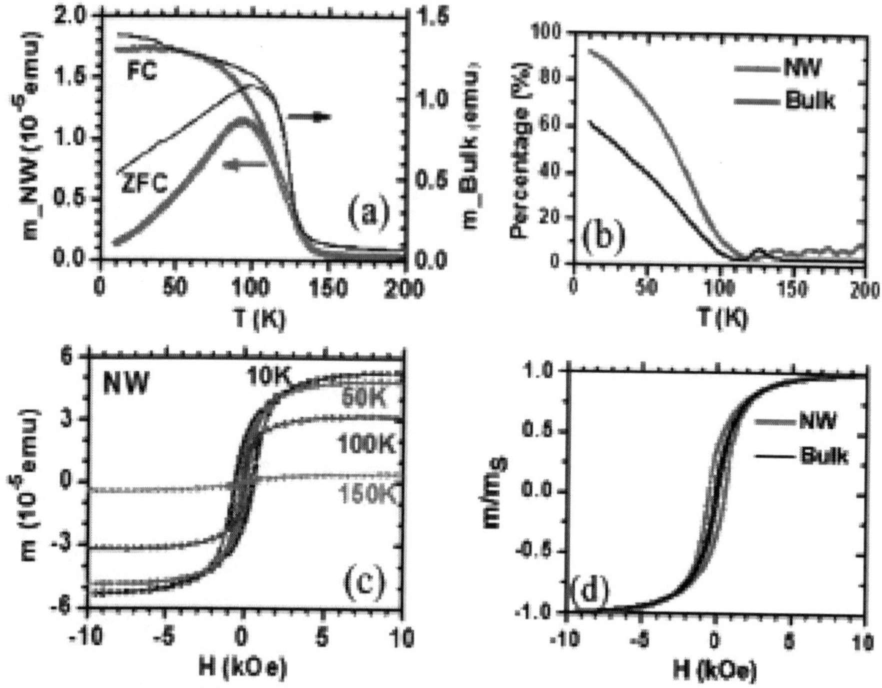

Figure 3: Magnetic measurements of LPCMO/MgO nanowires. (a)Magnetic moment versus temperature of the LPCMO/MgO nanowires (NW) and the LPCMO bulk polycrystalline sample after ZFC and FC [63]. The cooling field and the measuring field are both 200 Oe. (b) The percentage of the frozen phase defined as [*m*(FC)-*m*(ZFC)]/*m*(FC); (c) the field dependent magnetic moment of the LPCMO/MgO nanowires at different temperatures; and (d)the hysteresis loops of the nanowires and the bulk sample measured at $T=10$ K.

To better understanding the role of EPS in the emergent transport behaviors of manganite one-dimensional structures, some manganite wires are also fabricated from single thin film by optical lithography, where the width of the manganite wires is reduced to a scale on the order of the inherent phase separation. As one example, LPCMO is chosen as a representative model system due to its sub-micrometer scale phase separation that can be easily accessed by conventional lithographic fabrication processes. It is found that by reducing a single-crystal LPCMO thin film to a wire with a width comparable to a scale on the order of the inherent EPS, the system exhibits ultrasharp jumps in resistivity, as shown in Figure 4[27]. These jumps are attributed to a reduction of the transport lanes to a single channel. As the insulating barriers of the charge-ordered state are broken by the reduction of temperature or an increase in magnetic field, the resistance in the wire shows sharp jumps around the MIT, which reflects the nature of the first-order phase transition between ferromagnetic metal and charge-ordered insulator domains. Since the transport measurement can reveal the signature of phase transition of an individual EPS domain in a manganite wire, it becomes possible to probe the EPS domain dynamics of manganites. This is accomplished by setting an LPCMO wire at or very near the critical point of phase transition and measuring the phase fluctuation with high-time resolution. The very limited number of EPS domains that can be hosted in the wire effectively removes the problems associated with spatial averaging methods in the conventional transport measurements while allowing for a high temporal resolution. At the critical point of the MIT, single-domain fluctuations will show a clear signature in time-dependent resistivity measurements, as shown in Figure 5[29]. In Figure 5a, the resistivity ( ) of a 10 μm × 50 μm × 70 nm LPCMO wire under a 3.75-T magnetic field exhibits an ultrasharp jump at the MIT centered at 83 K. This is contrasted with the same sample in a film geometry (Figure 5a, inset) which shows a smooth transition from metallic to insulating behavior across a 150-K window. The extremely large jump in the resistivity of the wire results solely from its geometry's ability to remove the effects of spatial averaging in transport measurements. By setting the temperature of the LPCMO wire precisely in the middle of the 2-K window found in the temperature-dependent resistivity scan, it is possible to study the microscopic details of the transition in both space and time. Figure 5b shows the time-dependent resistivity while the wire is held at the transition temperature. It is clear that the apparent two-state system with resistivity jumps is actually comprised of a much richer multistate system. There are three inherent resistivity levels, each containing a further two-state fluctuation. However, in unconfined structures such as film (inset in Figure 5) or bulk counterpart, only white noise is observed. These data clearly show that the fluctuations that change the electrical resistance exist in these phase-separated manganite wires. It is observed that these fluctuations exist only near the transition temperature where electronic domains are fluctuating between FMM and COI and are not individually observable in films or bulk transport experiments. Therefore, the fluctuations in the wire are the direct signal of the microscopic fluctuations in EPS domains at the transition temperature. The comparable dimensions of the inherent domains to the wire result in a large change in the total wire resistance when a single domain fluctuates from one phase to another. Not only did these findings give us new insights into the mechanisms that drive electronic phase transitions, but they also open the door to engineering novel devices and could be applied as an on-chip digital randomizer as one example. Recently, large aspect-ratio (length-to-width >300) single-crystal nanowires of $La_{2/3}Ca_{1/3}MnO_3$ were also fabricated by combined optical and focused ion beam lithographies, which preserved their functional properties [66]. Remarkably, an enhanced magnetoresistance value of 34 % in an applied magnetic field of 0.1 T in the narrowest 150-nm nanowire was obtained. Such behavior is ascribed to the strain release at the edges together with a destabilization of the insulating regions. This opens new strategies to implement these structures in functional spintronic devices.

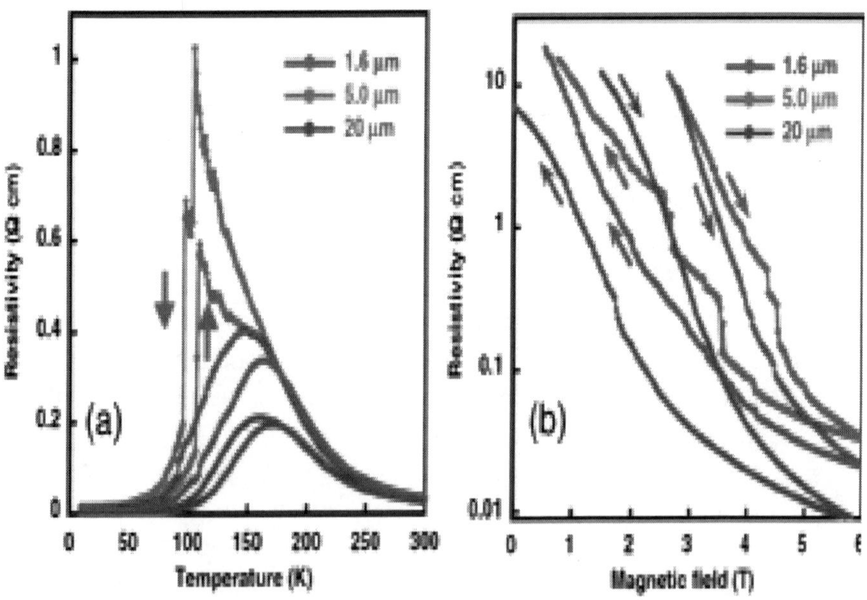

Figure 4: Resistivity versus temperature curves and resistivity vs. magnetic field curves. (a) Resistivity versus temperature (R-T) curves for the LPCMO wires under a 3.75-T magnetic field [27]. Arrows indicate the direction of the

temperature ramp. The R-T curves all exhibit hysteresis behavior in cooling-warming cycles, which is consistent with the coexistence of ferromagnetic metal and charge-ordered insulator domains in the LPCMO system. The MIT is rather smooth for both the 20-μm and the 5-μm wires. Ultrasharp and giant steps are clearly visible for the 1.6-μm wire. (b) Resistivity vs. magnetic field curves for the LPCMO wires measured at 110 K. Sudden step-like jumps are again visible in the 1.6-μm wire. Arrows indicate the sweeping directions of the magnetic field for each curve.

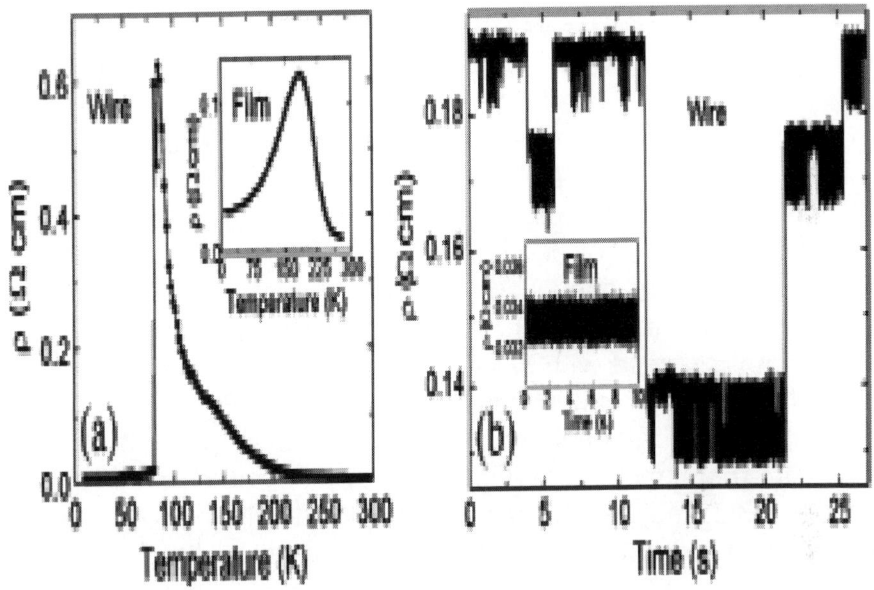

Figure 5: Time-dependent resistivity measurements. (a) Wire shows abrupt drop in resistivity at the MIT transition while the film shows a smooth transition (inset) [29]. (b) Resistivity of a wire when held at the transition temperature shows clear jumps associated with single electronic domain fluctuations. This behavior is not observed in the film, which only exhibits white noise (inset).

In addition to the manganite nanowires, the EPS in the manganite nanotubes are also investigated. Nanotubes are different from nanowires because they typically have a hollow cavity, whereas nanowires are completely filled with nanomaterials. From a viewpoint of device engineering and theoretical considerations, manganite nanotubes with nonplanar geometries have a variety of applications for nanoelectronic components [67-69], transducers, or sensors [70,71]. Recently, perovskite rare-earth manganese tubes such as $La_{0.67}Sr_{0.33}MnO_3$ (LSMO), $La_{0.67}Ca_{0.33}MnO_3$ (LCMO), and $La_{0.325}Pr_{0.300}Ca_{0.375}MnO_3$ (LPCMO) have been fabricated using a sol–gel template synthesis process [53,72,73]. Their typical length is about 6 to 8 μm and the average wall thickness is 45, 60, and 150 nm for LSMO, LCMO, and LPCMO, respectively [54]. The walls of the tubes are composed of magnetic nanograins, and their sizes are less than the critical size for multidomain formation in manganites. As a consequence, each particle that constitutes the nanotube walls is a single magnetic domain. Figure 6a shows the magnetizations of the LSMO, LCMO, and LPCMO nanotubes as a function of the temperature T measured at different applied magnetic fields (only show the data measured at $H=100$ Oe) following the next protocol: zero-field cooling (ZFC) (1 in Figure 6a), cooling the sample from the highest T with H=0 Oe; afterward, a magnetic field of H =100 Oe was applied and the magnetization data were collected increasing T. Field cool cooling (FCC) (2 in Figure 6a) is performed by measuring the magnetization by cooling the sample with H =100 Oe [54]. Finally, in field cool warming (FCW) (3 in the same plot), the system is warmed with H =100 Oe after FCC. It was noticed that there exists differences between the FCC (2*) and FCW (3*) curves in a broad temperature range for LPCMO nanotubes. Figure 6b displays the square-root temperature dependence of the coercive fields for the LCMO, LSMO, and LPCMO nanotubes [54]. Clearly, the coercive fields of the LCMO and LSMO nanotubes followed a linear dependence with the square root of temperature, whereas a nonlinear dependence was observed in LPCMO nanotubes, and the higher coercive field value was associated with the competition between the CO and the FM phases in the phase separated LPCMO nanotubes. Normally, a linear dependence is expected in the noninteracting particle systems, which can originate in the single magnetic domains that constitute the walls of the ferromagnetic nanotubes [74]. Therefore, as shown in Figure 6, the LSMO and LCMO nanotubes present a homogeneous ferromagnetic behavior below 340 and 258 K, respectively. The magnetic dead layer avoids the exchange interaction between the nanograins, but the dipolar interaction between them was detected which suggests a fanning array of magnetic moments along the tube axis. The coercive field temperature dependence indicates the presence of weak interactions. As for the LPCMO nanotubes, they became mainly ferromagnetic below 200 K. Their thermal hysteresis and the low magnetization values indicate the presence of an extra charge-ordered phase in the LPCMO nanotubes. The small grain size may generate small charge-disordered regions, which are easily transformed to a ferromagnetic region by field or increasing/decreasing temperature. Also, the charge-disordered phase attenuates the interaction between single magnetic domains when this phase is reduced by the application of a magnetic field; the system increases its ferromagnetic character. So, the control of the charge-disordered phase fraction could be used to tune the magnitude of the interaction between the single magnetic domains which affects the coercive fields.

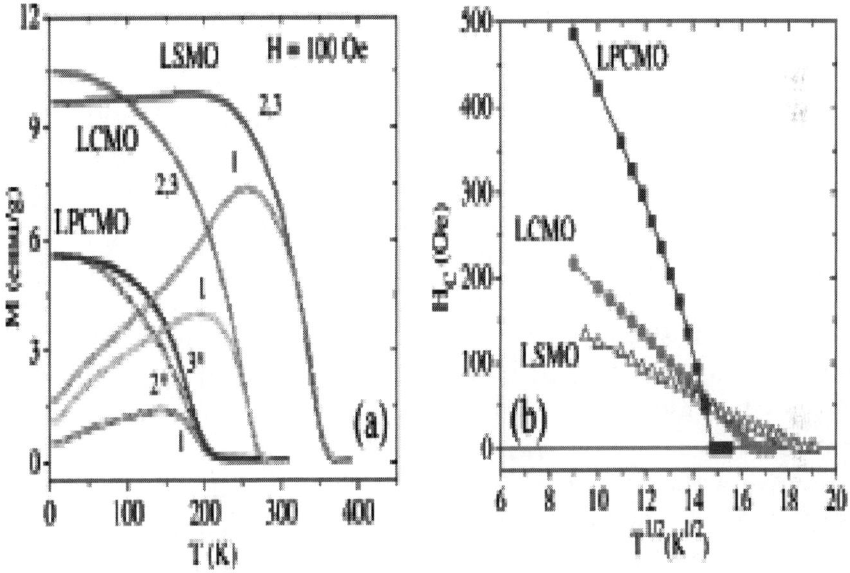

Figure 6: Magnetizations and square-root temperature dependence of the LSMO, LCMO, and LPCMO nanotubes. (a)*M* vs *T* at 100 Oe of LSMO, LCMO, and LPCMO nanotubes after different magnetothermal processes [54]. The numbers 1, 2, and 3 show the data collected in a 1 ZFC warming process after cooling with zero magnetic field, 2 FCC cooling process with a magnetic applied field of 100 Oe, and 3 FCW warming after the FCC process with 100 Oe. The asterisk indicates that the FCC and FCW in the LPCMO-nanotubes are different. (b) Square-root temperature dependence of the coercive fields for the LCMO, LSMO, and LPCMO nanotubes.

EPS in Manganite Nanostructured Films/ Patterns

In most CMR manganites, both the MIT and the amplitude of magnetoresistance are critically dependent upon the percolation of ferromagnetic metal domains in the system. Controlling the formation and the spatial distribution (size, density, symmetry, etc.) of the electronic domains will not only help to understand the origin of the EPS but also help to design manganites or other correlated electronic materials with desired properties for all-oxide-based electronic devices. Recently, a novel method called electronic nanofabrication (a conceptually new approach) is developed to control the formation and the spatial distribution of electronic domains in manganites [35]. In contrast to the conventional nanofabrication, the electronic nanofabrication patterns electronic states in materials without changing the actual size, shape, and chemical composition of the materials, which is a promising method for manganites. For example, magnetic Fe nanodots are grown on the surface of a 20-nm-thick $La_{0.7}Ca_{0.3}MnO_3/LaAlO_3(001)$ film, which could turn the film from an insulator to a metal with a high MIT temperature, as shown in Figure 7[75]. The underlying mechanism is understood to be the local magnetic exchange field between Fe and Mn spins that aligns the local Mn spins leading to the formation of a local metallic state. As shown in Figure 8, the MIT temperature can be also tuned by the density of Fe nanodots, which strongly indicates that the local metallic state follows the spatial locations of the Fe nanodots [75]. Besides the electronic nanofabrication technique, other methods such as atomic force microscopy lithography [28], electron-beam lithography (EBL) [76-80], focused ion beam (FIB) milling [33,34,81-84], and chemical growth and etching [85,86] are also used to fabricate manganite nanostructured patterns from oxide thin films. For example, microscopic $Pr_{0.65}(Ca_{0.75}Sr_{0.25})_{0.35}MnO_3$ (PCSMO) thin films were fabricated into patterns by EBL with width comparable to the length scale of EPS (~1 μm), where spontaneous resistance jumps along with the local Joule heating-induced step-like negative differential resistance were clearly observed [76]. Recently, LCMO microbridges with different widths were also fabricated by EBL method, where the MIT temperature was found to be decreased as reducing the bridge width, and the MIT even disappeared over the measured temperature range for the bridge with a width of 500 nm [76]. The underlying mechanism for this phenomenon is the confined geometry, which is dominated by the filamentary conduction mechanism. The magnetoresistance of the bridge also shows interesting behavior for enhanced *e-e* interactions in the presence of spin disorder; it can decrease and even change its sign in the bridges with widths of 1.5 and 1.0 μm under magnetic field of 1 T. The obvious size effects in the manganite microbridge nanopatterns are invaluable for further understanding the EPS phenomenon and its role in CMR effect.

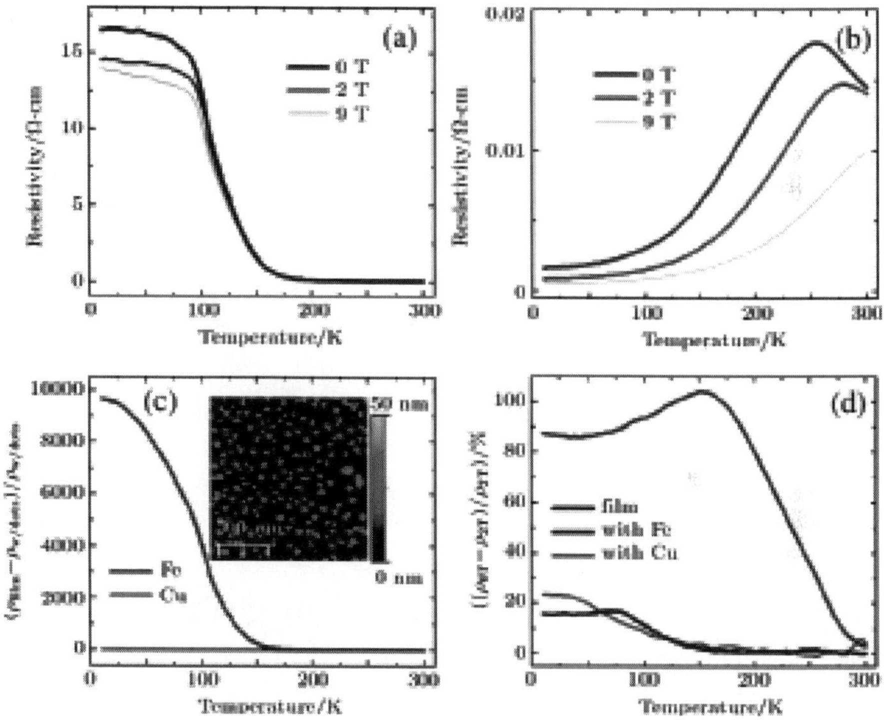

Figure 7: Transport properties of ultrathin LCMO film before and after application of nanodots [[75]]. (a) Resistivity behavior for 20-nm ultrathin film of $La_{0.7}Ca_{0.3}MnO_3$ showing insulating behavior and no clear metal-insulator transition. (b) Resistivity data of the same film after applying Fe nanodots to surface showing a recovery to bulk-like behavior with an MIT temperature of 255 K at 0 T (note change in scale). (c) Ferromagnetic Fe nanodots drive a huge change in the film's resistivity compared to the diamagnetic Cu nanodots. Insets: AFM images of typical nanodot coverages for Cu and Fe systems on LCMO films. (d)Magnetoresistive behavior shows a much higher magnetic response in the spin coupled system.

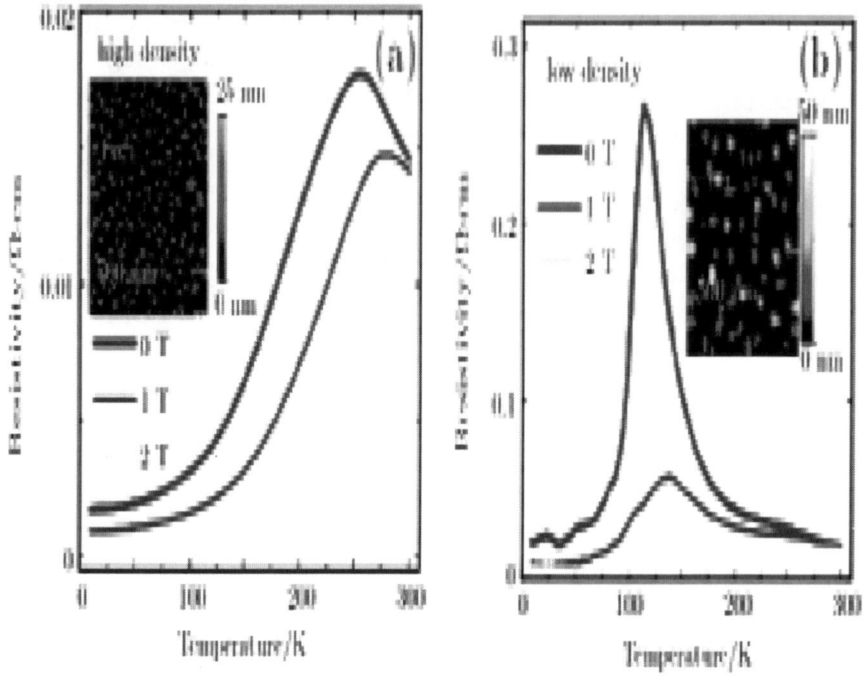

Figure 8: Comparison of transport properties with different Fe nanodot density. Resistive data for an ultrathin LCMO film after application of low density Fe nanodots shows recovery of the metal-insulator transition but with a much lower transition temperature than that seen at higher nanodot densities [75].

Origin of EPS in Perovskite Manganite Nanostructures

EPS as an inherent electronic inhomogeneity has been observed in real space with atomic-scale resolution in the perovskite manganites, which is generally regarded to be crucial for the CMR effect. This greatly stimulates a growing and theoretical interest in the EPS of perovskite manganite nanostructures. Now, the main theoretical approaches developed for investigating the EPS in perovskite manganite nanostructures can be classified into two categories, namely, approaches based on the model Hamiltonians and phenomenological theory. Dagotto and colleagues have developed one-orbital FM Kondo model and two-orbital model with Jahn-Teller phonons to investigate the EPS phenomenon in one-dimensional manganites [58,87-89]. In the model of Kondo lattice Hamiltonian with ferromagnetic Hund's coupling, the classical limit for the spin of the (localized) t_{2g} electrons is analyzed on lattices of dimension 1, 2, 3, and ∞ using several numerical methods [60,87,88]. They found that the expected double-exchange-induced strong tendencies to ferromagnetic correlations at low temperatures were in competition with a regime of phase separation, which occurred between the hole-undoped antiferromagnetic and hole-rich ferromagnetic regions. Although, the one-orbital model for manganites contains interesting physics, notably, a FM-AF competition that has similarities with those found in experiments. However, to explain the notorious orbital order tendency in Mn-oxides, it is crucial to use a model with two orbitals, where there is an electron Jahn-Teller phonon coupling and also Coulomb interactions [89,90]. Under the assumption that both localized e_g-spins and phonons are classical, the model without Coulombic terms can be studied fairly accurately using numerical and mean-field approximations. The calculated results for a one-dimensional system at low temperature by considering the two e_g orbitals and the Jahn-Teller phonons enrich the phase diagram considerably, as shown in Figure 9[90]. Obviously, the phase diagram is much rich, which includes different phases such as metallic and insulating regimes with orbital order. It is clear that phase separation appears at small e_g-densities between an electron-undoped AF-state and a metallic uniform-orbital-ordered FM state. Ahn et al. also proposed a model Hamiltonian including electron–phonon interactions and long-range elastic coupling between the local lattice distortions [91]. They presented a scenario for mesoscopic/microscopic inhomogeneities and suggested them to be the main source of the CMR effect. Since the physics of perovskite manganites is controlled by many degrees of freedom at the atomic level and the associated energy scales, Ramakrishnan et al. also developed a microscopic model for manganites that includes all the important energy scales present in them [92]. In this model, the degeneracy of the two e_g orbitals is split into two types of states, *l* and *b* at each manganese site by electron-lattice coupling. The *l* state is polaronic and has an exponentially reduced hopping amplitude, whereas the electrons within the *b* states hop with the bare amplitude. Such two-fluid model of manganites demonstrates colossal magnetoresistive response and reproduces the physical transport properties confirmed by the experimental measurements. Due to the local strong Mott-

Hubbard repulsion, the simultaneous occupation at both the *l* and *b* states on a given site is not allowed; this model exhibits macroscopic phase separation, where the region with *l* polarons corresponds to the charge-ordered states, while that with *b* electrons corresponds to the FM metal. However, due to the frustration of the long-range Coulomb interactions, the phase separation usually exhibits as a structure of nanoscopic FM metallic domains with interspersed domains of polaron Coulomb glass, similar to the charge-ordered FM state revealed by the electron holography [11]. Therefore, the manganites are intrinsically inhomogeneous at length scales of nanometers due to the strong electronic correlations. A phenomenological Ginzburg-Landau theory approach is also developed by using a Landau free-energy function and introducing the term of electronic softness to rationalize the possibility of phase coexistence and electronic inhomogeneities [93]. In this approach, magnetic and charge modulations are argued to coexist in new thermodynamic phases in contrast to the previous models where the phase separation originates from disorder or as a strain-induced kinetic phenomenon. This approach leads to a rich diagram of equilibrium phases, qualitatively similar to those seen experimentally. The success of this approach argues for a fundamental reinterpretation of the nature of charge modulation in manganite materials, from a localized to a more extended 'charge-density wave' picture. The same symmetry considerations that favor textured coexistence of charge and magnetic order may apply to many electronic systems with competing phases. The resulting 'electronically soft' phases of matter with incommensurate, inhomogeneous, and mixed order may be general phenomena in correlated systems.

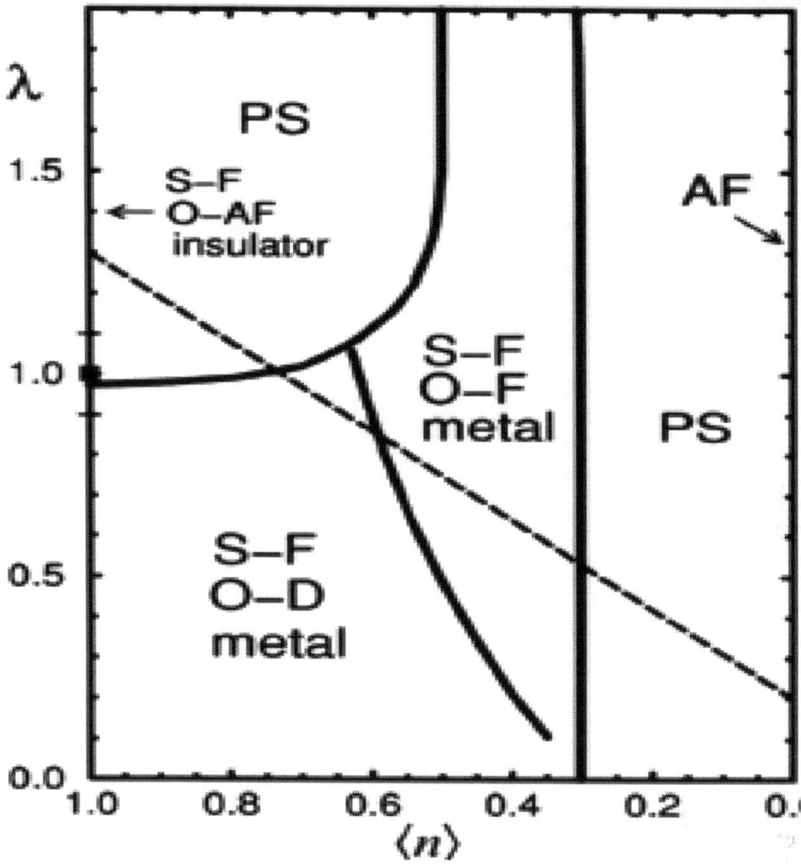

Figure 9: Phase diagram of two-orbital model in one-dimensional and T ~0 including Jahn-Teller phonons, obtained with Monte Carlo techniques [[90]]. S-F labels a spin-ferromagnetic configuration. O-F, O-AF, and O-D denote a state where the orbital degrees of freedom are ordered uniformly, staggered or they are disordered, respectively; PS indicates a phase separated state, and AF in an antiferromagnetic state. The Hund-coupling is $J_H = 8$, the Heisenberg coupling between localized classical spins $J_{AF} = 0.05$, both in units of the hopping amount the same orbitals.

Since a number of competing energy scales are operative in manganite oxides giving rise to a large number of electronic orders such as spin, charge, and orbital (and associated lattice order), the emergence of these orders and the associated couplings between them should be considered in a full Hamiltonian model for manganites, which makes the theoretical understanding of the EPS quite complex. Much work is further needed in this challenging area of research.

CONCLUSIONS

In recent years, a remarkable progress has been achieved in understanding the EPS phenomenon in low-dimensional perovskite manganite nanostructures such as manganite nanoparticles, nanowires, nanotubes, and nanostructured films/patterns. This progress is mainly made possible by building upon the experimental measurements and theoretical approaches, and clearly establishes the phase completion as the main source of the CMR effect in manganite oxides. The shape and scale of EPS are different for different systems with electronic domain sizes ranging from a few nanometers to several micrometers. However, the microscopic nature of the EPS in perovskite manganite oxides and the physical origin of the large length-scales that arise in these systems are still unclear although much theoretical work based on model Hamiltonians and phenomenological models has been done. Clearly, much further work is needed to be done on both the experimental and theoretical fronts to understand the nature of the EPS manganite oxides, especially at the nanoscale. On the experimental side, a new technique is needed to be developed to control the formation and the spatial distribution of electronic domains in manganite oxides, which should allow to simultaneously probe EPS domains with different electronic states and give the vital information on phase formation, movement, and fluctuation. Such a novel technique called electronic nanofabrication has been developed. In striking contrast to the conventional nanofabrication, the electronic nanofabrication patterns electronic states in materials without changing the actual size, shape, and chemical composition of the materials, which allows one to control the global physical properties of the system at a very fundamental level and greatly enhances the potential for realizing true oxide electronics. The theorists need to quantitatively clarify the electronic properties of the various manganite phases based on microscopic Hamiltonians, including strong electron–phonon Jahn-Teller and/or Coulomb interactions. Thus, quantitative calculations for addressing the CMR effect help us to better understand the physical nature of EPS phenomenon. However, to get a full understanding of the EPS phenomenon in low-dimensional manganite nanostructures, much work remains to be done for realizing its practical applications in oxide electronics.

AUTHORS' CONTRIBUTIONS

LL and XZ designed the structure and modified the manuscript articles; LL drafted the manuscript. HW participated in the sequence alignment. All authors read and approved the final manuscript.

ACKNOWLEDGEMENTS

This work was partially supported by National Natural Science Foundation of China (Grant Nos. 11174122 and 11134004), National Basic Research Program of China (Grant Nos. 2009CB929503 and 2012CB619400), and the open project from National Laboratory of Solid State Microstructures, Nanjing University.

REFERENCES

1. Schiffer P, Ramirez AP, Bao W, Cheong SW: Low temperature magnetoresistance and the magnetic phase diagram of $La_{1-x}Ca_xMnO_3$.*Phys Rev Lett* 1995, 75:3336.
2. Salamon MB, Jaime M: The physics of manganites: structure and transport.*Rev Mod Phys* 2001, 73:583.
3. Dagotto E: Complexity in strongly correlated electronic systems. *Science* 2005, 309:257.
4. Zhang L, Israel C, Biswas A, Greene RL, de Lozanne A: Direct observation of percolation in a manganite thin film.*Science* 2002, 298:805.
5. Uehara M, Mori S, Chen CH, Cheong SW: Percolative phase separation underlies colossal magnetoresistance in mixed-valent manganites.*Nature* 1999, 399:560.
6. Asamitsu A, Tomioka Y, Kuwahara H, Tokura Y: Current switching of resistive states in magnetoresistive manganites.*Nature* 1997, 388:50.
7. Mathur ND, Littlewood PB: Mesoscopic texture in manganites. *Phys. Today* 2003, 56:25.
8. Rao CNR, Kundu AK, Seikh MM, Sudheendra L: Electronic phase separation in transition metal oxide systems.*Dalton Trans* 2004, 19:3003.

9. Dagotto E, Hotta T, Moreo A: Colossal magnetoresistant materials: the key role of phase separation.*Phys Rep* 2001, 344:1.
10. Shenoy VB, Gupta T, Krishnamurthy HR, Ramakrishnan TV: Coulomb interactions and nanoscale electronic inhomogeneities in manganites.*Phys Rev Lett* 2007, 98:097201.
11. LoudonJC, MathurND, MidgleyPA: Charge-orderedferromagnetic phase in La0.5Ca0.5MnO3.*Nature* 2002, 420:797.
12. Ma JX, Gillaspie DT, Plummer EW, Shen J: Visualization of localized holes in manganite thin films with atomic resolution. *Phys Rev Lett* 2005, 95:237210.
13. Tao J, Niebieskikwiat D, Varela M, Luo W, Schofield MA, Zhu Y, Salamon MB, Zuo JM, Pantelides ST, Pennycook SJ: Direct imaging of nanoscale phase separation in la0.55ca0.45mno3: relationship to colossal magnetoresistance.*Phys Rev Lett* 2009, 103:097202.
14. Murakami Y, Kasai H, Kim JJ, Mamishin S, Shindo D, Mori S, Tonomura A: Ferromagnetic domain nucleation and growth in colossal magnetoresistive manganite.*Nat Nanotech* 2010, 5:37.
15. Lai KJ, Nakamura M, Kundhikanjana W, Kawasaki M, Tokura Y, Kelly MA, Shen ZX:Mesoscopic percolating resistance network in a strained manganite thin film.*Science* 2010, 329:190.
16. Shenoy VB, Sarma DD, Rao CNR: Electronic phase separation in correlated oxides: the phenomenon, its present status and future prospects.*ChemPhysChem* 2006, 7:2053.
17. Shenoy VB, Rao CNR: Electronic phase separation and other novel phenomena and properties exhibited by mixed-valent rare-earth manganites and related materials.*Phil Trans R Soc A* 2008, 366:63. Rao SS, Anuradha KN, Sarangi S, Bhat SV: Weakening of charge order and antiferromagnetic to ferromagnetic switch over in Pr0.5Ca0.5MnO3 nanowires.*Appl Phys Lett* 2005, 87:182503.

18. Rao SS, Tripathi S, Pandey D, Bhat SV: Suppression of charge order, disappearance of antiferromagnetism, and emergence of ferromagnetism in $Nd_{0.5}Ca_{0.5}MnO_3$nanoparticles.*Phys Rev B* 2006, 74:144416.
19. Sarkar T, Ghosh B, Raychaudhuri AK, Chatterji T: Crystal structure and physical properties of half-doped manganite nanocrystals of less than 100-nm size.*Phys Rev B* 2008, 77:235112.

20. Zhang T, Dressel M: Grain-size effects on the charge ordering and exchange bias in $Pr_{0.5}Ca_{0.5}MnO_3$: The role of spin configuration. *Phys Rev B* 2009, 80:014435.
21. Jirák Z, Hadová E, Kaman O, Knížek K, Maryško M, Pollert E, Dlouhá M, Vratislav S:Ferromagnetism versus charge ordering in the $Pr_{0.5}Ca_{0.5}MnO_3$ and $La_{0.5}Ca_{0.5}MnO_3$nanocrystals.*Phys Rev B* 2010, 81:024403.
22. Markovich V, Fita I, Wisniewski A, Jung G, Mogilyansky D, Puzniak R, Titelman L, Gorodetsky G: Spin-glass-like properties of $La_{0.8}Ca_{0.2}MnO_3$ nanoparticles ensembles. *Phys Rev B* 2010, 81:134440.
23. Markovich V, Fita I, Wisniewski A, Mogilyansky D, Puzniak R, Titelman L, Gorodetsky G: Size-driven magnetic transitions in $La_{1/3}Ca_{2/3}MnO_3$ nanoparticles.*J Appl Phys* 2010, 108:063918.
24. Huang XH, Ding JF, Zhang GQ, Hou Y, Yao YP, Li XG: Size-dependent exchange bias in $La_{0.25}Ca_{0.75}MnO_3$ nanoparticles.*Phys Rev B* 2008, 78:224408.
25. Markovich V, Fita I, Wisniewski A, Mogilyansky D, Puzniak R, Titelman L, Martin C, Gorodetsky G: Size effect on the magnetic properties of antiferromagnetic $La_{0.2}Ca_{0.8}MnO_3$nanoparticles. *Phys Rev B* 2010, 81:094428.
26. Zhai HY, Ma JX, Gillaspie DT, Zhang XG, Ward TZ, Plummer EW, Shen J: Giant discrete steps in metal-insulator transition in perovskite manganite wires.*Phys Rev Lett* 2006, 97:167201.
27. Yanagisawa Y, Tanaka H, Kawai T, Pellegrino L: Digitalized magnetoresistance observed in (La, Pr, Ca)MnO3 nanochannel structures.*Appl. Phys. Lett.* 2006, 89:253121.
28. Ward TZ, Zhang XG, Yin LF, Zhang XQ, Liu M, Snijders PC, Jesse S, Plummer EW, Cheng ZH, Dagotto E, Shen J: Time-resolved electronic phase transitions in manganites.*Phys Rev Lett* 2009, 102:087201.
29. Ward TZ, Gai Z, Guo HW, Yin LF, Shen J: Dynamics of a first-order electronic phase transition in manganites.*Phys Rev B* 2011, 83:125125.
30. Ward TZ, Liang S, Fuchigami K, Yin LF, Dagotto E, Plummer EW, Shen J: Reemergent metal-insulator transitions in manganites exposed with spatial confinement.*Phys Rev Lett* 2008, 100:247204.

31. Wu JC, Sun H, Da HX, Li ZY: Study of anomalous I-V characteristics in spatially confined manganite structures.*Appl Phys Lett* 2007, 91:102501.

32. Singh-Bhalla G, Selcuk S, Dhakal T, Biswas A, Hebard AF: Intrinsic tunneling in phase separated manganites.*Phys Rev Lett* 2009, 102:077205.

33. Singh-Bhalla G, Biswas A, Hebard AF: Tunneling magnetoresistance in phase-separated manganite nanobridges.*Phys Rev B* 2009, 80:144410.

34. Shen J, Ward TZ, Yin LF: Emergent phenomena in manganites under spatial confinement.*Chin Phys B* 2013, 22:017501.

35. Wollan EO, Koehler WC: Neutron diffraction study of the magnetic properties of the series of perovskite-type compounds [(1 - x)La, xCa]MnO_3.*Phys Rev* 1955, 100:545.

36. Jonker GH, Van Santen JH: Ferromagnetic compounds of manganese with perovskite structure.*Physica* 1950, 16:337.

37. Van Santen JH, Jonker GH: Electrical conductivity of ferromagnetic compounds of manganese with perovskite structure.*Physica* 1950, 16:599.

38. Zener C: Interaction between the d-shells in the transition metals. *Phys Rev* 1951, 81:440.

39. Searle CW, Wang ST: Studies of the ionic ferromagnet (LaPb) MnO_3. V. Electric transport and ferromagnetic properties.*Can J Phys* 1970, 48:2023.

40. Jin S, Tiefel TH, McCormack M, Fastnacht RA, Ramesh R, Chen LH: Thousandfold change in resistivity in magnetoresistive La-Ca-Mn-O films.*Science* 1994, 264:413.

41. Woodward PM, Cox DE, Vogt T, Rao CNR, Cheetham AK: Effect of compositional fluctuations on the phase transitions in $(Nd_{1/2}Sr_{1/2})MnO_3$.*Chem Mater* 1999, 11:3528.

42. Fäth M, Freisem S, Menovsky AA, Tomioka Y, Aarts J, Mydosh JA: Spatially inhomogeneous metal-insulator transition in doped manganites.*Science* 1999, 285:1540.

43. Mori S, Chen CH, Cheong SW: Pairing of charge-ordered stripes in (La, Ca)MnO3.*Nature* 1998, 392:473. Renner C, Aeppli G, Kim BG, Soh YA, Cheong SW: Atomic-scale images of charge ordering in a mixed-valence manganite.*Nature* 2002, 416:518.

44. De Teresa JM, Ibarra MR, Algarabel PA, Ritter C, Marquina C, Blasco J, Garcia J, del Moral A, Arnold Z: Evidence for magnetic polarons in the magnetroresistive perovskites.*Nature* 1997, 386:256.
45. Zhang T, Zhou TF, Qian T, Li XG: Particle size effects on interplay between charge ordering and magnetic properties in nanosized $La_{0.25}Ca_{0.75}MnO_3$.*Phys Rev B* 2007, 76:174415.
46. Zhu D, Zhu H, Zhang Y: Hydrothermal synthesis of La0.5Ba0.5MnO3 nanowires.*Appl Phys Lett* 2002, 80:1634.
47. Zhang T, Jin CG, Qian T, Lu XL, Bai JM, Li XG: Hydrothermal synthesis of single-crystalline $La_{0.5}Ca_{0.5}MnO_3$ nanowires at low temperature.*J Mater Chem* 2004, 14:2787.
48. Li D, Wang Y, Xia Y: Electrospinning of polymeric and ceramic nanofibers as uniaxially aligned arrays.*Nano Lett* 2003, 3:1167.
49. Jugdersuren B, Kang S, DiPietro RS, Heiman D, McKeown D, Pegg IL, Philip J: Large low field magnetoresistance in La0.67Sr0.33MnO3 nanowire devices.*J Appl Phys* 2011, 109:016109.
50. Shantha K, Raychaudhuri AK: Growth of an ordered array of oriented manganite nanowires in alumina templates.*Nanotechnol* 2004, 15:1312.
51. Shankar KS, Kar S, Raychaudhuri AK, Subbanna GN: Fabrication of ordered array of nanowires of La0.67Ca0.33MnO3 (x=0.33) in alumina templates with enhanced ferromagnetic transition temperature.*Appl Phys Lett* 2004, 84:993.
52. Curiale J, Sánchez RD, Troiani HE, Ramos CA, Pastoriza H, Leyva AG, Levy P: Magnetism of manganite nanotubes constituted by assembled nanoparticles.*Phys Rev B* 2007, 75:224410.
53. Freeman MR, Choi BC: Advances in magnetic microscopy. *Science* 2001, 294:1484.
54. Markovich V, Puzniak R, Mogilyansky D, Wu XD, Suzuki K, Fita I, Wisniewski A, Chen SJ, Gorodetsky G: Exchange bias effect in $La_{0.2}Ca_{0.8}MnO_3$ antiferromagnetic nanoparticles with two ferromagnetic-like contributions.*J Phys Chem C* 2011, 115:1582.
55. Zhang T, Wang XP, Fang QF: Evolution of the electronic phase separation with magnetic field in bulk and nanometer $Pr_{0.67}Ca_{0.33}MnO_3$ particles.*J Phys Chem C* 2011, 115:19482.

56. Kodama RH, Berkowitz AE, McNiff EJ Jr, Foner S: Surface spin disorder in $NiFe_2O_4$nanoparticles.*Phys Rev Lett* 1996, 77:394. Wu JH, Lin JG: Study on the phase separation of La0.7Sr0.3MnO3 nanoparticles by electron magnetic resonance.*J Magn Magn Mater* 2006, 304:e7.
57. Yunoki S, Hu J, Malvezzi AL, Moreo A, Furukawa N, Dagotto E: Phase separation in electronic models for manganites.*Phys Rev Lett* 1998, 80:845.
58. Han S, Li C, Liu ZQ, Lei B, Zhang DH, Jin W, Liu XL, Tang T, Zhou CW: Transition metal oxide core-shell nanowires: generic synthesis and transport studies.*Nano Lett* 2004, 4:1241.
59. Nagashima K, Yanagida T, Tanaka H, Seki S, Saeki A, Tagawa S, Kawai T: Effect of the heterointerface on transport properties of in situ formed Mgo/titanate heterostructured nanowires.*J Am Chem Soc* 2008, 130:5378.
60. Li L, Li H, Zhai XF, Zeng CG: Fabrication and magnetic properties of single-crystalline La0.33Pr0.34Ca0.33MnO3/MgO nanowires. *Appl Phys Lett* 2013, 103:113101.
61. Ghivelder L, Parisi F: Dynamic phase separation in $La_{5/8-y}Pr_yCa_{3/8}MnO_3$.*Phys Rev B* 2005, 71:184425.
62. Niebieskikwiat D, Sanchez RD: Pinning of elastic ferromagnetic/antiferromagnetic interfaces in phase-separated manganites.*J Phys Condens Matter* 2012, 24:436001.
63. Marín L, Morellón L, Algarabel PA, Rodríguez LA, Magén C, De Teresa JM, Ibarra MR:Enhanced magnetotransport in nanopatterned manganite nanowires.*Nano Lett* 2014, 14:423.
64. Postma HWC, Teepen T, Yao Z, Grifoni M, Dekker C: Carbon nanotube single-electron transistors at room temperature.*Science* 2001, 293:76.
65. Wolf SA, Awschalom DD, Buhrman RA, Daughton JM, Von Molnár S, Roukes ML, Chtchelkanova AY, Treger DM: Spintronics: a spin-based electronics vision for the future.*Science* 2001, 294:1488-1495.
66. Tseng GY, Ellenbogen JC: Toward nanocomputers.*Science* 2001, 294:1293.
67. Bachtold A, Hadley P, Nakanishi T, Dekker C: Logic circuits with carbon nanotube transistors.*Science* 2001, 294:1317.

68. Hueso L, Mathur ND: Nanotechnology: dreams of a hollow future.*Nature* 2004, 427:301.
69. Levy P, Leyva AG, Troiani H, Sánchez RD: Nanotubes of rare-earth manganese oxide.*Appl Phys Lett* 2003, 83:5247.
70. Leyva AG, Stoliar P, Rosenbusch M, Lorenzo V, Levy P, Albonetti C, Cavallini M, Biscarini F, Troiani HE, Curiale J, Sánchez RD: Microwave assisted synthesis of manganese mixed oxide nanostructures using plastic templates.*J Solid State Chem* 2004, 177:3949.
71. Cullity BD: *Introduction to Magnetic Materials*. Reading, USA: Addison-Wesley; 1972.
72. Ward TZ, Gai Z, Xu XY, Guo HW, Yin LF, Shen J: Tuning the metal-insulator transition in manganite films through surface exchange coupling with magnetic nanodots.*Phys Rev Lett* 2011, 106:157207.
73. Wu T, Mitchnell JF: Creation and annihilation of conducting filaments in mesoscopic manganite structures.*Phys Rev B* 2006, 74:214423.
74. Guo X, Li PG, Wang X, Fu XL, Chen LM, Lei M, Zheng W, Tang WH: Anomalous positive magnetoresistance effect in $La_{0.67}Ca_{0.33}MnO_3$ microbridges.*J Alloy Compd* 2009, 485:802.
75. Arnal T, Khvalkovskii AV, Bibes M, Mercey B, Lecoeur P, Haghiri-Gosnet AM: Electronic properties of domain walls in $La_{2/3}Sr_{1/3}MnO_3$: magnetotransport measurements on a nanopatterned device.*Phys Rev B* 2007, 75:220409.
76. Beekman C, Zaanen J, Aarts J: Nonlinear mesoscopic transport in a strongly cooperative electron system: The $La_{0.67}Ca_{0.33}MnO_3$ microbridge.*Phys Rev B* 2011, 83:235128.
77. Beekman C, Komissarov I, Aarts J: Large electric-field effects on the resistance of $La_{0.67}Ca_{0.33}MnO_3$ microstructures.*Phys Rev B* 2012, 85:245115.
78. Pallecchi I, Pellegrino L, Caviglia A, Bellingeri E, Canu G, Gazzadi GC, Siri AS, Marré D:Current-driven hysteresis effects in manganite spintronics devices.*Phys Rev B* 2006, 74:014434.
79. Pallecchi I, Gadaleta A, Pellegrino L, Gazzadi GC, Bellingeri E, Siri AS, Marré D: Probing of micromagnetic configuration in manganite channels by transport measurements.*Phys Rev B* 2007, 76:174401.

80. Ruotolo A, Oropallo A, Miletto Granozio F, Pepe GP, Perna P, Uccio USD, Pullini D: Current-induced domain wall depinning and magnetoresistance in La0.7Sr0.3MnO3 planar spin valves. *Appl Phys Lett* 2007, 91:132502.

81. Céspedes O, Watts SM, Coey JMD, Dörr K, Ziese M: Magnetoresistance and electrical hysteresis in stable half-metallic La0.7Sr0.3MnO3 and Fe3O4 nanoconstructions.*Appl Phys Lett* 2005, 87:083102.

82. Bhalla GS: *Size Effects in Phase Separated Manganite Nanostructures*. Florida, USA: Ph. D.Thesis. University of Florida; 2009.

83. Nakajima T, Tsuchiya T, Ueda Y, Kumagai T: Probing electronic-phase-separated insulating domains in the metallic phase of patterned perovskite manganite microwires.*Phys Rev B* 2009, 80:020401.

84. Dagotto E, Yunoki S, Malvezzi AL, Moreo A, Hu J, Capponi S, Poilblanc D, Furukawa N:Ferromagnetic Kondo model for manganites: phase diagram, charge segregation, and influence of quantum localized spins.*Phys Rev B* 1998, 58:6414.

85. Dagotto E: *Nanoscale Phase Separation and Colossal Magnetoresistance*. Berlin, Germany: Springer; 2003.

86. Yunoki S, Moreo A, Dagotto E: Phase separation induced by orbital degrees of freedom in models for manganites with Jahn-Teller phonons.*Phys Rev Lett* 1998, 81:5612.

87. Moreo A, Yunoki S, Dagotto E: Phase separation scenario for manganese oxides and related materials.*Science* 1999, 283:2034.

88. Ahn KH, Lookman T, Bishop AR: Strain-induced metal–insulator phase coexistence in perovskite manganites.*Nature* 2004, 428:401.

89. Ramakrishnan TV, Krishnamurthy HR, Hassan SR, Pai GV: Theory of insulator metal transition and colossal magnetoresistance in doped manganites.*Phys Rev Lett* 2004, 92:157203.

90. Milward GC, Calderon MJ, Littlewood PB: Electronically soft phases in manganites. *Nature* 2005, 433:607.

Chapter 6

Optical Interactions in Plasmonic Nanostructures

Wounjhang Park

Department of Electrical, Computer & Energy Engineering, University of Colorado, Boulder 80309-0425, CO, USA

ABSTRACT

We present a review of plasmonic nanostructures in which the constituent materials are coupled together by optical interactions. The review first provides a comprehensive coverage of theoretical framework where the optical interactions are described by the multiple scattering among the nanostructures. We then discuss the two limiting cases of weak and strong interactions. The weak interaction regime is described well by the effective medium theory while the strong interaction regime requires theoretical tools that can describe the new eigenmodes delocalized over the entire system. Weakly interacting plasmonic nanostructures have been studied extensively in the metamaterials research, which has been a major research thrust in photonics during the past decade. This review covers some of the latest examples exhibiting perfect absorption and invisibility. Strongly coupled systems started to receive attention recently. As a representative example, plasmonic molecules exhibiting Fano resonance are discussed in detail. Plasmonic nanostructures offer an excellent platform to engineer nanoscale optical fields. With the recent progress in nanofabrication technologies, plasmonic nanostructures offer a highly promising pathway to discovering new phenomena and developing novel optical devices.

INTRODUCTION

In the Drude theory, metal is modeled as a system of free electron gas undergoing forced oscillation with some damping. Other than some notable exceptions such as the strong interband absorption involving the *d* band in transition metals, the Drude model generally provides good descriptions of the optical properties of metals. It predicts that in the low frequency region the induced polarization oscillates 180° out-of-phase with the external field, resulting in negative permittivity and consequently high reflectivity. At certain frequencies, the electron gas can also undergo collective oscillations whose quantum is called plasmon. The collective oscillation may involve volume charge density (volume plasmon) or surface charge density (surface plasmon). Volume plasmons are longitudinal oscillations and cannot be excited by electromagnetic waves which are normally transverse. Surface plasmons, however, interact strongly with light. By storing a significant portion of its energy in oscillating electron gas, surface plasmon offers an effective means to localize light down to length scales much smaller than wavelength. For this reason, the past decades have seen an explosive growth in research on surface plasmon nanophotonics.

One of the topics of high interest is nanoscale waveguiding with potential applications in integrated photonics for which a large body of literature exists [1]. Another important class of plasmonic nanostructures more relevant to this review are those supporting localized surface plasmon (LSP) modes. These structures include metal nanoparticles of various shapes, nanoholes in metal films, and some variations of these. The LSP resonances typically manifest themselves with strong scattering and absorption. The characteristic color arising from the LSP resonances has enabled a wide range of applications from the medieval stained glasses to modern optical sensors. For example, colorimetric sensing of biomolecules and ions has been demonstrated [2,3]. Also, strongly scattering nanoparticles have been used to enhance contrast in dark-field microscopy and optical coherence tomography [4] while thermal ablation sing strongly absorbing nanoparticles is being actively developed for targeted therapies of various diseases [5,6].

LSP naturally provides a highly localized and intense optical field which consequently enhances a variety of optical processes. The best-known example is surface enhanced Raman scattering (SERS). Using a rough silver surface, Raman scattering by a single molecule has been observed with enhancements up to a factor of 10^{14}[7]. Much of the enhancement is believed to arise from the local field enhancement due to the highly localized hot spots produced by the silver nanostructure [8]. Surface plasmon resonance can also be used to enhance luminescence [9] and Förster energy transfer process [10, 11]. Since the optical processes depend on the local field strength, strong emphasis has been placed on achieving higher local field enhancement using novel nanostructure geometries. These structures are generally termed nano-antennas analogous to conventional antennas in the microwave frequency region and have been the subject of extensive research in the past decade [12]. Since hot spots are typically formed in the nano-gap between plasmonic nanoparticles, the nano-antenna research naturally directed attention to the interaction between LSP modes, which is the main subject of this review. An important recent development in this field is the concept of plasmon hybridization [13]. In this picture, the individual LSP modes in closely spaced nanoparticles hybridize to form a set of new modes delocalized over the entire structure, in much the same way atomic orbitals hybridize to form molecular orbitals in real molecules. For this reason, these plasmonic nanostructures are also called plasmonic molecules. The hybridized modes in plasmonic molecules have been extensively studied. The simplest form of plasmonic molecule is a dimer or a pair of nanoparticles in which the plasmon resonance of single nanoparticle hybridizes to produce symmetric and anti-symmetric dimer plasmons [14]. Hybridization in more complex structures has also been studied, e.g. trimer [15], quadrumer [16], hexamer [17] and most widely heptamer [18-22]. In these complex structures, symmetry properties of the hybridized plasmon modes should be described by the group representation theory [23]. In this formalism, how each plasmon mode transforms under the symmetry operations is described by the irreducible representations of the symmetry group. This allows for a systematic description of modes with complex field profiles and also their interaction with incident optical fields, providing a complete picture when combined with the numerical techniques that can precisely calculate the hybridized LSP modes. It is noted that the group theoretical approach classifying the symmetry properties of plasmon modes does not require any approximations. For example, in nanostructures with features sizes much smaller than the wavelength, the plasmon modes exhibit predominantly electric dipole nature and are often approximated as point dipoles. The group theoretical approach, however, does not require dipole approximation and is equally applicable to larger and more complex structures exhibiting higher order mulitpole components.

The strength and nature of the optical interaction between plasmonic nanostructures varies with the distance. For long distances, the interaction is mainly radiative and involves propagating electromagnetic waves undergoing multiple scattering among the plasmonic nanostructures. In this case, the original character of the individual LSP modes should largely be preserved. For short distances, evanescently decaying near-field components should be taken into account. In principle, the near-field interaction can be treated the same way as the far-field interaction in the general framework of multiple scattering theory. That is, the evanescent modes undergo multiple scattering between nanostructures as well as the propagating modes. In extreme short distances, where the individual LSP modes have significant overlaps with one another, the near-field interaction will dominate and hybridization resulting in fully delocalized modes would occur. This review is aimed at providing a unified view of plasmonic nanostructures in which the strength of the interaction among individual metallic components is varied. We will first present a theoretical framework followed by examples of nanostructures and their optical properties in the weak and strong coupling regimes.

REVIEW

Theoretical Framework

Classical Description of Surface Plasmon Resonance

Surface plasmon, while inherently a quantum mechanical entity, is generally well described by the classical theory where surface plasmon presents itself as a surface resonance. There are of course quantum mechanical characteristics that cannot be fully described by the classical theory such as nonlocal response of electrons [24,25], tunneling and screening effect [26], spontaneous and stimulated emission of photons [27] and plasmons [28,29], to name a few. Nevertheless, the classical theory has proved sufficient for a wide range of systems. The classical theory is based on the Maxwell's equations where the boundary conditions are set up by the material properties specified by the macroscopic permittivity and permeability. For the planar geometry, surface wave solutions with fields decaying exponentially away from the interfaces can be found. In the simplest example of a single planar interface between semi-infinite metal and dielectric, the surface wave vector, *q*, is given by the dispersion relation [30],

$$q = \frac{\omega}{c}\sqrt{\frac{\varepsilon_1 \varepsilon_2}{\varepsilon_1 + \varepsilon_2}} \tag{1}$$

where ε_1 and ε_2 are the permittivities of metal and dielectric medium, respectively, and often Drude model is used for the metal permittivity, $\varepsilon_1(\omega)=1+\omega_p^2/(\omega^2+i\gamma\omega)$ where ω_p is the plasma frequency and γ is the damping parameter. This surface wave solution specifies the optical mode often called surface plasmon-polariton (SPP). For the lossless case where the permittivities are purely real, the dispersion relation has a singularity when $\varepsilon_1+\varepsilon_2=0$. This condition defines the surface plasmon frequency, ω_s, which is given by $\omega_s=\omega_p/\sqrt{1+E_2}$, when using the Drude model for metal permittivity. Here ω_p is the plasma frequency of the electron gas. When there is finite loss, the metal permittivity is complex and the resonance condition should be written as which defines the condition for maximum electric field amplitude. This theory can be straightforwardly extended to more complex geometries involving multiple interfaces to obtain SPP solutions [31-34]. The localized surface plasmon modes (LSPs) generally require numerical calculations as the surface plasmon modes are highly sensitive to the details of the given geometry. An important exception is the highly symmetric geometry of sphere and circular cylinder for which a complete analytical solution can be obtained by using the Mie theory where the scattering and absorption cross-sections are given by the scattering coefficients *an* and b_n corresponding to various multipole terms [35]. For small spheres, the scattering coefficients *an* have vanishing denominators when $m^2=-(n+1)/n$, where m is the relative refractive index of sphere relative to the embedded medium and n is an integer. This leads to a sharp increase in both scattering and absorption. In the limit of extremely small sphere, the extinction is dominated by the leading term, a_1, resulting in the resonance condition, $m^2=-2$ or $\varepsilon_1=-2\varepsilon_2$ when using the Drude model for the metal permittivity. For the lossless case where the permittivities are purely real, the resonance condition is satisfied when $\omega=\omega_p/\sqrt{2+E_2}$, which is sometimes called the Fröhlich frequency. Thus, the LSPs supported by metallic spheres and cylinders may be considered special cases of Mie resonances.

Multiple Scattering Theory

We now consider a cluster of nanoparticles or nanostructures which individually support LSP resonances. The key additional consideration needed to properly describe the cluster is the multiple scattering between the individual components. Since analytical solutions are available for spheres, it is instructive to consider a cluster of nanospheres. In this case, the scattering by the individual spheres can be described by the standard Mie theory. The fundamental difference from the single sphere case is that for each sphere in the cluster the incident field is the combination of the actual incident field impinging on the cluster and the scattered fields by all other spheres in the cluster [36-38]. To briefly provide an overview of this process, we first remind the reader of the standard Mie theory for a single sphere where all fields are described in the coordinate system with the sphere located at the origin. The incident and scattered fields are then expressed in terms of the vector spherical harmonics,

$$
\begin{aligned}
\mathbf{E}_{scat} &= \sum_{n=1}^{\infty}\sum_{m=-n}^{n} iE_{mn}\left(a_{mn}\mathbf{N}_{mn}^{(3)} + b_{mn}\mathbf{M}_{mn}^{(3)}\right)\\
\mathbf{E}_{int} &= \sum_{n=1}^{\infty}\sum_{m=-n}^{n} iE_{mn}\left(d_{mn}\mathbf{N}_{mn}^{(1)} + c_{mn}\mathbf{M}_{mn}^{(1)}\right)\\
\mathbf{E}_{inc} &= \sum_{n=1}^{\infty}\sum_{m=-n}^{n} iE_{mn}\left(p_{mn}\mathbf{N}_{mn}^{(1)} + q_{mn}\mathbf{M}_{mn}^{(1)}\right)
\end{aligned}
\tag{2}
$$

where E_{scat}, E_{int} and E_{inc} represent scattered, internal and incident fields, respectively, M_{mn} and N_{mn} are the vector spherical harmonics. Similar expressions for the magnetic fields are readily obtained from the Maxwell's equations. The coefficients p_{mn} and q_{mn} are naturally dependent on the exact form of the incident field. For plane wave incidence, all terms with $m \neq 1$ vanish and the full expressions for the coefficients, a_n, b_n, c_n, d_n, p_n and q_n are available in the literature [39]. To describe multiple scattering in a cluster made of L spheres, the coefficients, p_{mn} and q_{mn}, describing the incident field for a sphere needs to be rewritten to include the scattered fields by all other spheres. This in turn requires translating all vector harmonics functions separately obtained in the coordinate systems of individual spheres into the common coordinate system so they may be added. This is achieved by the addition theorem of vector spherical harmonics:

$$\mathbf{M}_{mn} = \sum_{\nu=0}^{\infty}\sum_{\mu=-\nu}^{\nu}\left(A0_{\mu\nu}^{mn}\mathbf{M}'_{mn} + B0_{\mu\nu}^{mn}\mathbf{N}'_{mn}\right)$$
$$\mathbf{N}_{mn} = \sum_{\nu=0}^{\infty}\sum_{\mu=-\nu}^{\nu}\left(B0_{\mu\nu}^{mn}\mathbf{M}'_{mn} + A0_{\mu\nu}^{mn}\mathbf{N}'_{mn}\right) \quad (3)$$

where $\mathbf{M}_{mn}$ and $\mathbf{N}_{mn}$ are the vector spherical harmonics about origin O, M'_{mn} and N'_{mn} are the vector spherical harmonics about origin O′, and AO_{uv}^{mn} and are the additional coefficients whose complete expressions are available in the literature [36]. One can then express the scattered field by the *l*th sphere in the coordinate system of *j*th sphere using the new coefficients, $p_{mn}lj$ and $q_{mn}lj$,

$$p_{mn}^{lj} = -\sum_{\nu=0}^{\infty}\sum_{\mu=-\nu}^{\nu}\left(a_{\mu\nu}^{l}A_{mn}^{\mu\nu}(l,j) + b_{\mu\nu}^{l}B_{mn}^{\mu\nu}(l,j)\right)$$
$$q_{mn}^{lj} = -\sum_{\nu=0}^{\infty}\sum_{\mu=-\nu}^{\nu}\left(a_{\mu\nu}^{l}B_{mn}^{\mu\nu}(l,j) + b_{\mu\nu}^{l}A_{mn}^{\mu\nu}(l,j)\right) \quad (4)$$

where A_{uv}^{mn} and B_{uv}^{mn} are related to AO_{uv}^{mn} and $B0_{\mu\nu}{}^{mn}$ as given in Ref. [36]. Imposing the boundary conditions at the surfaces of the spheres finally leads to a set of linear equations

$$a_{mn}^{j} = a_{n}^{j}\left\{p_{mn}^{jj} - \sum_{l\neq j}\sum_{\nu=0}^{\infty}\sum_{\mu=-\nu}^{\nu}\left(a_{\mu\nu}^{l}A_{mn}^{\mu\nu}(l,j) + b_{\mu\nu}^{l}B_{mn}^{\mu\nu}(l,j)\right)\right\}$$
$$b_{mn}^{j} = b_{n}^{j}\left\{q_{mn}^{jj} - \sum_{l\neq j}\sum_{\nu=0}^{\infty}\sum_{\mu=-\nu}^{\nu}\left(a_{\mu\nu}^{l}B_{mn}^{\mu\nu}(l,j) + b_{\mu\nu}^{l}A_{mn}^{\mu\nu}(l,j)\right)\right\} \quad (5)$$

Once the scattering coefficients, and, are found from the above equations, then the total scattered field by the entire cluster can be obtained by using the addition theorem one more time to express all fields in the common coordinate system of j_0th sphere,

$$\mathbf{E}_{scat} = \sum_{n=1}^{\infty} \sum_{m=-n}^{n} iE_{mn}\left(a_{mn}\mathbf{N}_{mn}^{(3)} + b_{mn}\mathbf{M}_{mn}^{(3)}\right) \tag{6}$$

Where

$$\begin{aligned} a_{mn} &= \sum_{l=1}^{L}\sum_{\nu=0}^{\infty}\sum_{\mu=-\nu}^{\nu}\left(a_{\mu\nu}^{l}A_{mn}^{\mu\nu}(l,j_0) + b_{\mu\nu}^{l}B_{mn}^{\mu\nu}(l,j_0)\right) \\ b_{mn} &= \sum_{l=1}^{L}\sum_{\nu=0}^{\infty}\sum_{\mu=-\nu}^{\nu}\left(a_{\mu\nu}^{l}B_{mn}^{\mu\nu}(l,j_0) + b_{\mu\nu}^{l}A_{mn}^{\mu\nu}(l,j_0)\right) \end{aligned} \tag{7}$$

The scattering cross section can be written in a similar fashion to the single sphere Mie scattering case,

$$C_{scat} = \frac{4\pi}{k^2}\sum_{n=0}^{\infty}\sum_{m=-n}^{n} n(n+1)(2n+1)\frac{(n-m)!}{(n+m)!}\left(\left|a_{mn}\right|^2 + \left|b_{mn}\right|^2\right) \tag{8}$$

A similar multiple scattering theory has also been developed for an infinitely periodic array of spheres [40-42]. In this case, since the system has infinite extent, the main concern is the propagation of light inside the periodic array. Due to the similarity to the Kohn-Korringa-Rostocker (KKR) method widely used for electronic energy band structure studies, this technique is sometimes called the vector KKR method. Once again, vector spherical harmonics are used to take advantage of the spherical geometry. Following the original development by Stefanou et al.[40,43], we begin with the general expression for spherical wave expansion given in equation (6) for a two-dimensional (2D) periodic array of spheres. When a plane wave is incident on the 2D array, the scattered wave must be expressed as a sum of the waves scattered by all spheres in the 2D plane. Due to the periodicity, the wave scattered by a sphere at $r=R_n$ differs only by a simple phase factor, $\exp(i k_{||} \cdot R n)$, from the wave scattered by a sphere at the origin. Here, $\mathbf{k}_{||}$ represents the tangential component of k-vector parallel to the plane of spheres. The scattered field can then be written as,

$$E_{scat} = \sum_{n=1}^{\infty} \sum_{m=-n}^{n} \left(\frac{i}{k} b_{mn}^{E} \nabla \times \sum_{R_n} \exp(i\mathbf{k}_{\|} \cdot \mathbf{R}_n) h_n(kr_n) \mathbf{X}_{mn}(\theta, \phi) + b_{mn}^{H} \sum_{R_n} \exp(i\mathbf{k}_{\|} \cdot \mathbf{R}_n) h_n(kr_n) \mathbf{X}_{mn}(\theta, \phi) \right) \tag{9}$$

where b_{mn} are coefficients to be determined, $r_n = r - R_n$, h_n is the spherical Hankel function and X_{mn} is defined as $\sqrt{n(n+1)}\, X_{mn} = -ir \times \nabla Y_{mn}$ where Y_{mn} is the spherical harmonics function. While we try to keep the notation of original authors as much as possible, we note that the above equation is essentially in the same form as equation (6) where b^E_{mn} and b^H_{mn} in equation (9) are analogous to a_{mn} and b_{mn} coefficients in equation (6). The phase factor, $\exp(i k_{||} \cdot R_n)$, in the above equation allows us to express the scattered field in terms of plane waves with wave vectors,

$$K_g^{\pm} = \left(\mathbf{k}_{\|} + \mathbf{g}, \pm\sqrt{k^2 - (\mathbf{k}_{\|} + \mathbf{g})^2} \right) \tag{10}$$

Here $\mathbf{k}_{||}$ represents the reduced wave vector defined as $\mathbf{k}_{tangential} = \mathbf{k}_{||} + \mathbf{g}$, where $\mathbf{g}$ is the reciprocal lattice vectors of the 2D lattice, so that $\mathbf{k}_{||}$ now represents a vector within the unit cell of the 2D reciprocal lattice or the surface Brillouin zone. The sign in the superscript indicates forward and backward propagating waves in the direction perpendicular to the 2D plane. Now we construct the same plane wave expansion for the incident wave as well and obtain a matrix equation relating the coefficients of the incident wave and the total scattered wave,

$$[E_{scat}]^{s}_{g,i} \equiv \sum_{g',i'} M^{ss'}_{gi,g'i'} [E_{inc}]^{s'}_{g',i'} \tag{11}$$

where the new subscripts *i, i'* indicate the Cartesian components. The explicit expression for the matrix M′ may be found in Ref. [44]. Equation (11) fully describes the interaction of a 2D periodic array of spheres with an incident light but, since the summation is infinite, actual numerical evaluations will involve truncation which introduces numerical errors. Now we extend the theory to three-dimensional (3D) array composed of multiple layers of 2D arrays by constructing a transfer matrix describing the multiple scattering effects between the adjacent 2D planes. That is, the field between the *n*th and $(n+1)$ th layers are determined by the scattering by the two neighboring planes of spheres. More specifically, the backward propagating wave in the region between *n*th and $(n+1)$th layer is the sum of the forward propagating wave in the same region back-scattered by $(n+1)$th layer and the backward propagating wave between the $(n+1)$th and $(n+2)$ th layers transmitted (or forward-scattered) by the $(n+1)$th layer. The forward propagating wave is similarly defined. The transmission and reflection matrices are obtained from the matrix **M** in equation (11) modified to account for the shift of origin between the two adjacent planes. This way, one obtains equations relating fields at adjacent layers which must also satisfy the Bloch theorem because the system is periodic in the 3rd dimension as well. Combining the multiple scattering equations and the Bloch theorem, one obtains an eigenvalue equation,

$$\begin{pmatrix} \mathbf{Q}^{++} & \mathbf{Q}^{+-} \\ -[\mathbf{Q}^{--}]^{-1}\mathbf{Q}^{-+}\mathbf{Q}^{++} & [\mathbf{Q}^{--}]^{-1}[\mathbf{Q}^{-+}\mathbf{Q}^{+-}] \end{pmatrix} \begin{pmatrix} \mathbf{E}^{+}(n) \\ \mathbf{E}^{-}(n+1) \end{pmatrix} = \exp\left(i\mathbf{k}\times\mathbf{a}_3\right)\begin{pmatrix} \mathbf{E}^{+}(n) \\ \mathbf{E}^{-}(n+1) \end{pmatrix} \tag{12}$$

Here the matrices Q are the scattering matrices defined as $Q^{ss'}_{gg'} = M^{ss'}_{gg'} \exp[i(sK^{s}_{g}.d_r + s'K^{s'}_{g'}.d_l)]$ where s and s′ are + or – signs, g and g′ index the 2D reciprocal lattice vectors, d_r and d_l are right and left translation vectors, the matrix M is given in equation (11), and $\mathbf{a}_3$ is the primitive unit vector representing the periodicity in the 3rd dimension. It is noted that this last part is the standard transfer matrix formalism and can be used for non-spherical scatterers as well [45]. For given ω and $k_{||}$ in the surface Brillouin zone, the matrix in equation (12) is fully specified and thus can be solved using the standard matrix inversion techniques. The plane wave expansion over the reciprocal lattice vectors of course will need to be truncated for numerical evaluations.

Effective Medium Theory

Now we explore the limiting cases where the multiple scattering theory presented above can either be simplified significantly or is inadequate. An obvious possibility of simplification is found when the interaction among the scatterers is weak. In this case, the multiple scattering effect will be small and may be ignored in the extreme case. A commonly employed approximation for such a case is the effective medium theory which attempts to obtain an average response of a non-uniform medium composed of multiple constituents. The most widely used effective medium theory is the Maxwell Garnett theory [46], which is derived from the Lorentz local field relation covered in many textbooks. Briefly, Lorentz considered an imaginary spherical cavity in a dielectric medium and stated that the local field is simply given by the sum of the dielectric response of the medium and the additional fields due to the surface charge on the spherical cavity and due to the atoms inside the cavity. The lattermost contribution cancels out to zero in a highly symmetric crystal structure. This leads to the famous expression,

$$\mathbf{E}_{local} = \mathbf{E} + \frac{4\pi\mathbf{P}}{3} \tag{13}$$

where **E** and **P** are the macroscopic field and polarization of the medium. Relating the local field to the induced atomic dipole moment, $\mathbf{p} = \alpha\mathbf{E}_{local}$, we arrive at the Clausius-Mossotti equation,

$$\frac{\varepsilon - 1}{\varepsilon + 2} = \frac{4\pi N a}{3} \tag{14}$$

where ε is the permittivity of the medium and N and α are atomic density and polarizability, respectively. Maxwell Garnett effective medium theory simply replaces the atomic polarizability in the Clausius-Mossotti equation with the polarizability of a spherical object. In the electrostatic limit, the polarizability of a sphere in vacuum has a simple form,

$$\alpha = \left(\frac{\varepsilon_s - 1}{\varepsilon_s + 2}\right) a^3 \tag{15}$$

where *a* is the radius of the spherical object and ε_s is the permittivity of the sphere. The Maxwell Garnett mixing rule is obtained by substituting equation (15) into (14),

$$\frac{\varepsilon - 1}{\varepsilon + 2} = f_s \cdot \frac{\varepsilon_s - 1}{\varepsilon_s + 2} \tag{16}$$

where f_s is the volume fraction of the sphere. For spheres embedded in a dielectric matrix with permittivity ε_m, we simply replace 1 and 2 in equation (16) with ε_m and $2\varepsilon_m$, respectively.

We now make a connection with the more rigorous multiple scattering theory presented earlier. In the Mie theory for a single spherical scatterer, the scattering coefficients *an* and *bn* are generally expressed as infinite series of spherical Hankel functions. Using the power series expansions of the spherical Hankel functions, it can be shown that the leading terms of these coefficients are $a_1 \sim x^3, b_1 \sim x^5$,

$a_2 \sim x^5$, $b_2 \sim x^7$, etc., where $x = 2\pi a/\lambda$. Therefore, for small spheres, we may retain only the leading term in a_1. Since the scattering and absorption cross sections are proportional to $|a_1|^2$and **Im**(a_1) (imaginary part of a_1), respectively, the scattering and absorption efficiencies defined as the ratio of scattering and absorption cross sections to the physical cross section depend on $x^4$and x, respectively, yielding the well-known Rayleigh scattering results. More specifically, the leading term of coefficient a_1 is

$$a_1 = -\frac{i2x^3}{3} \cdot \frac{\varepsilon_s - \varepsilon_m}{\varepsilon_s + 2\varepsilon_m} \tag{17}$$

The relationship between the a_1 coefficient and the polarizability is apparent,

$$\alpha = \frac{3ia^3}{2x^3} a_1 \tag{18}$$

This relationship reveals one of the key approximations involved in the Maxwell Garnet effective medium theory: retaining only the leading term of the leading coefficient a_1 in the general solution of the Mie theory. This is essentially the electrostatic approximation in which the polarizability of small spheres is modeled by the static polarizability given in equation (15). A modest improvement on this approximation may be achieved by the extended Maxwell Garnett theory in which the full expansion of a_1 is used to calculate the polarizability [47]. The extended Maxwell Garnett theory has been used successfully for metamaterials [48,49]. However, if higher order terms in coefficienta_1 are to be included, it is logical to include b_1 and a_2 terms as well. These terms represent the contributions of the magnetic dipole and electric quadrupole terms, respectively. The inclusion ofb_1 can be done straightforwardly by defining effective permeability in exactly the same way effective permittivity is calculated except that a_1 is replaced by b_1 to define magnetic polarizability. It has been shown that the optical properties of small particle composites may exhibit magnetic permeability even when the constituent materials are all non-magnetic [50]. Thus, the use of extended Maxwell Garnett theory is more appropriate when both effective permittivity and permeability are considered. It has also been attempted to improve the Maxwell Garnett theory using dressed polarizability to better account for the resonant effect [51]. In addition to neglecting higher multipole contributions to the light scattering and absorption, which is appropriate for small particles, the Maxwell Garnett effective medium theory also ignores multiple scattering among particles. This can be seen from the facts that the polarizability is defined by the single sphere Mie coefficient and that the effective permittivity is constructed based on the Lorentz local field concept. There are other issues related to the Maxwell Garnett effective medium theory such as the Lorentz catastrophe and non-symmetry and there exist other effective medium theories that improve on those shortcomings [52]. However, it can be said in general that the effective medium theory is appropriate for small particles with low volume fraction where both the higher order multipole contributions and inter-particle interaction by multiple scattering can be ignored.

Plasmon Hybridization

We now us consider the opposite extreme where the particles are closely spaced and thus strongly interacting with one another. In this case, especially near plasmon resonances, strongly overlapping evanescent fields exist, leading to significant modification in the nature of the resonance. In principle, this near-field interaction can be described by the multiple scattering theory presented earlier. However, there is a practical difficulty in that it requires the inclusion of a large number partial waves and thus makes the numerical computation extremely challenging. Instead it is more profitable to look for new eigenmode solutions described by the collective modes delocalized over the entire structure. Prodan et al. proposed the plasmon hybridization model [13]. To be sure, the concept of plasmon hybridization itself is not new. A classic example is coupling of two surface plasmon modes in a metal thin film and the resultant formation of symmetric and anti-symmetric modes [53,54]. But Prodan's model is highly useful as it provides a complete set of eigenmode solutions in complex nanostructures. The plasmon hybridization model considers the electron gas as incompressible fluid subject to Coulombic interaction. The plasmon resonance is then represented by self-sustained oscillations of the electron fluid which may be obtained from the Euler-Lagrange equation for the Lagrangian,

$$L = \frac{n_0 m}{2} \int \eta \nabla \eta dS - \frac{1}{2} \int \frac{\sigma(\mathbf{r})\sigma(\mathbf{r}')}{\mathbf{r}-\mathbf{r}'} dSdS' \tag{19}$$

where n_0 and m are the electron density and mass, respectively, and σ is the surface charge density. This technique was shown to agree well with the full electrodynamic simulation results in the electrostatic limit where all of the involved length scales are small compared to the wavelength. It was successful in describing the plasmonic resonances in complex structures including nanoshell, nanoegg and nanorice [55]. The model has also been applied to multi-particle systems such as dimer, trimer, quadrumer, hexamer and heptamer [14,20,56]. A similar method has been proposed by Mayergoyz et al. who derived a boundary integral eigenvalue equation using the electrostatic approximation. In this approach, the surface plasmon resonance is described as an eigenstate of the homogeneous boundary integral equation [57],

$$\sigma(\mathbf{r}) = \frac{\gamma}{2\pi} \int \sigma(r') \frac{(\mathbf{r}\text{-}\mathbf{r}')}{|\mathbf{r}\text{-}\mathbf{r}'|^3} \cdot \hat{\mathbf{n}} dS \tag{20}$$

For an arbitrary geometry, the boundary integral equation (20) can be solved numerically to obtain the eigenvalue γ and eigenfunction $\sigma(\mathbf{r})$. The eigenvalue γ is related to the permittivity $\varepsilon(\omega)$ of the metal as

$$\mathrm{Re}\varepsilon(\omega) = \varepsilon_b \left(\frac{1+\gamma}{1-\gamma} \right) \tag{21}$$

where ε_b is the permittivity of the background material. It is thus possible to determine the resonance frequency from the permittivity of the metal. The eigenfunction $\sigma(\mathbf{r})$ describes the self-sustained surface charge distribution. This technique has been used to describe collective plasmon resonance supported by hexamers and heptamers made of gold nanoparticles and nanorods[17,21,22].

It should be noted that both of the two techniques described above employ electrostatic approximation and thus cannot describe the retardation effect. Their applicability is therefore limited to extremely small particles or low frequencies. Nevertheless, these methods proved highly useful in the exploration of coupled plasmon modes in complex nanostructures consisting of multiple nanoparticles. The main reason is the applicability of elegant group theoretical method in a fashion similar to the molecular orbital theory. Just as the molecular orbital can be constructed from the symmetry adapted linear combination of atomic orbitals, the hybridized plasmon modes can be constructed from the symmetry adapted linear combination of individual nanoparticle plasmon modes. This enables a highly effective and systematic exploration of plasmon modes in a complex, strongly coupled nanostructure. Here we briefly describe this approach using the boundary integral method for a collection of nanoparticles. First, the plasmon modes of an isolated nanoparticle can be found from the eigenvalue equation given in (20). For the collection of N nanoparticles, the surface charge distribution can be expressed as a linear combination of the individual nanoparticle eigenmodes,

$$\sigma(\mathbf{r}) = \sum_{\alpha=1}^{N} \sum_{j} a_{\alpha}^{j} \sigma_{\alpha}^{j}(\mathbf{r}) \tag{22}$$

Using equation (20), we may write

$$\sigma(\mathbf{r}) = \frac{\Lambda}{2\pi}\sum_{\beta,k}\int a_\beta^k \sigma_\beta^k(\mathbf{r}_\beta)\frac{(\mathbf{r}-\mathbf{r}_\beta)}{|\mathbf{r}-\mathbf{r}_\beta|^3}\cdot\hat{\mathbf{n}}dS \tag{23}$$

where σ_β^k represents the surface charge distribution of kth mode of βth nanoparticle. One can then use the bi-orthogonality between the surface charge distribution, $\sigma(\mathbf{r})$, and the surface dipole distribution, $\tau(\mathbf{r})$, to derive,

$$\begin{aligned} a_n^m &= \frac{\Lambda}{2\pi}\sum_{\beta,k} a_\beta^k\left\{\int \tau_\alpha^m(\mathbf{r})\hat{\mathbf{n}}\cdot\int \sigma_\beta^k(\mathbf{r}_\beta)\frac{(\mathbf{r}-\mathbf{r}_\beta)}{|\mathbf{r}-\mathbf{r}_\beta|^3}dS_\beta dS\right\} \\ &= \frac{\Lambda}{2\pi}\sum_{\beta,k} K_{\alpha\beta}^{mk}a_\beta^k \end{aligned} \tag{24}$$

The total surface charge distribution in equation (22) must satisfy the boundary integral eigenvalue equation in (20) where the integral is conducted over the surfaces of all nanoparticles. However, the problem now has been transformed to a matrix eigenvalue equation given in (24). Physically, the matrix K represents the interaction between the surface dipole density of the mth mode of αth particle and the charge distribution of the kth mode of βth particle. Now suppose the arrangement of the nanoparticles is symmetric and certain symmetry operations keep the nanoparticle cluster unchanged. This means the symmetry operators leave the interaction matrix K invariant. Then, the symmetry operators and matrix K commute and consequently they share the same eigenvectors. It is now possible to classify all eigenvectors of matrix K based on how they transform under various symmetry operations. In the language of group theory, the eigenvectors are classified by the irreducible representations of the symmetry group. These eigenvectors possess well-defined transformation characteristics under all symmetry operators that leave the structure invariant and form the symmetry-adapted basis set. The symmetry properties of the plasmon modes proved powerful in analyzing the interaction between various plasmon modes and also between light and plasmons. For example, the incident light interacts with the surface plasmon mode via the induced dipole moment. For a given symmetry group, the dipole operator belongs to a certain irreducible representation. Thus, it naturally follows that only those surface plasmon modes belonging to the same irreducible representation as the dipole operator can interact with light. Other modes are the so-called dark modes that do not interact with incident light and thus cannot be detected by experiments. Furthermore, surface plasmon modes belonging to different irreducible representations do not mix together as their basis functions are mutually orthogonal. These properties provide powerful insights into how complex plasmon modes evolve and interact in complex nanostructures.

Weakly Coupled Plasmonic Nanostructures

As discussed in the previous section, when the plasmonic nanostructures are only weakly coupled, they may be described well by the effective medium theory. There is a large body of literature on metamaterials which are composed of deep sub-wavelength scale structures and are generally described well by the effective medium theory. The initial development of metamaterials was driven largely by the negative index materials for which excellent reviews are available [58,59]. In this section, we focus on the latest development of plasmonic nanostructures in the weak coupling regime.

Perfect Absorber

An interesting application of plasmonic nanostructures is the perfect absorber which absorbs all incident light with minimal reflection and transmission. Such a material would find a wide range of applications in, for example, coatings for optical instruments, sensors, and photothermal or photovoltaic energy conversion devices. An inspection of Fresnel's coefficients immediately reveals that perfect absorption is achieved when the real part of the refractive index is matched with the incident medium and the imaginary part is small. This condition would result in negligibly small reflection and, if the material is thick enough, vanishing transmission as well. This is, in fact, how perfectly absorbing metamaterial was obtained using aligned carbon nanotube arrays [60]. This example illustrates clearly that a perfect absorber does not have to involve a plasmonic nanostructure [60,61]. It is, however, of great technological importance to achieve perfect absorption in thin films and plasmonic nanostructures are ideally suited for that purpose.

One of the simplest examples is a line grating. It is well known that a line grating on a metal surface can couple the free space light into surface plasmons. An ideal grating presents infinitesimal modulation of surface so that the plasmon modes of the original flat surface are well preserved and the grating only provides the extra momentum for coupling. However, it was found that, even in a grating as shallow as 20 nm, additional surface plasmon modes strongly localized at the grating lines exist and interact with the surface plasmon-polariton modes. Lu et al. investigated silver nanograting structures fabricated on the back-electrode of a PCDTBT:PCBM organic photovoltaic device [62]. The system supports two surface plasmon modes when excited by transverse-magnetic (TM, electric field perpendicular to the grating lines) polarized incident light. One is the localized surface plasmon (LSP) mode whose field profile is strongly localized at the grating lines and wavelength insensitive to the incident angle and active layer thickness. The other is the surface plasmon-polariton (SPP) mode strongly interacting with the grating period and thus highly sensitive to incident angle and active layer thickness. Figure 1(a) shows the wavelength of surface plasmon induced absorption peaks as a function of active layer thickness, revealing two branches with distinct dependence on active layer thickness. In general, as the active layer thickness is increased, the refractive index of the dielectric side of the surface is increased, resulting in a red shift in surface plasmon wavelength. The surface plasmon dispersion relation for a dielectric loaded metal surface can be calculated numerically [34]. As shown by the light blue crosses in Figure 1(a), the surface plasmon wavelength exhibits a red shift with increasing active layer thickness and this line coincides well with the SPP branch of the silver nanograting overcoated with PCDTBT:PCBM active layer. This indicates that the 20 nm tall silver nanograting acts as an ideal grating for the SPP modes which retain the characteristics of those in the flat dielectric loaded metal surface. The other branch representing LSP modes is unique to the grating structure and not observed in flat interfaces. Due to the strongly localized nature, it is insensitive to the thickness of the dielectric layer. For thicknesses near 30 nm, the two branches exhibit anti-crossing, indicative of strong mixing between the LSP and SPP modes. The field profiles in Figure 1(b) and (c) show the two contrasting patterns for LSP and SPP modes while those in Figure 1(d) and (e) reveal considerable mixing. Since these modes strongly affect the absorption spectra, the knowledge of the existence of two different modes and their interaction is essential to designing light trapping plasmonic grating structures for photovoltaic devices.

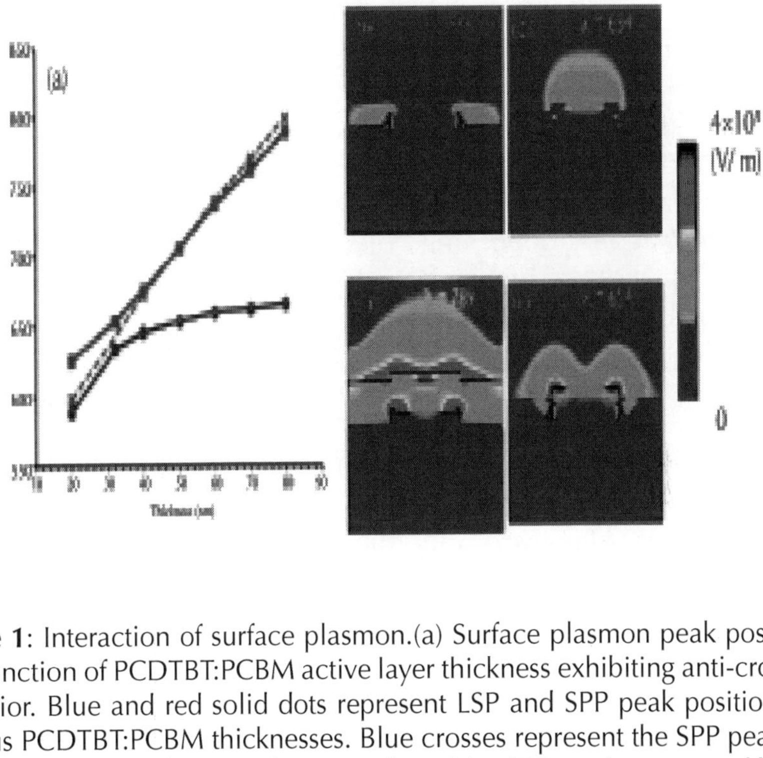

Figure 1: Interaction of surface plasmon.(a) Surface plasmon peak positions as a function of PCDTBT:PCBM active layer thickness exhibiting anti-crossing behavior. Blue and red solid dots represent LSP and SPP peak positions for various PCDTBT:PCBM thicknesses. Blue crosses represent the SPP peak positions calculated from the effective index of the SPP mode supported by the flat silver-PCDTBT:PCBM-air multilayer structure. (b, c) Electric field profiles for the surface plasmon modes in the 80 nm thick PCDTBT:PCBM on silver grating and (d, e) electric field profile for the surface plasmon modes in the 30 nm thick PCDTBT:PCBM on silver grating. To clarify the structure the bottom grating is the silver nanograting fabricated by nanoimprint lithography. The top surface is also corrugated simply by the result of depositing on a corrugated surface (From Ref. [62]).

Interaction of surface plasmon.(a) Surface plasmon peak positions as a function of PCDTBT:PCBM active layer thickness exhibiting anti-crossing behavior. Blue and red solid dots represent LSP and SPP peak positions for various PCDTBT:PCBM thicknesses. Blue crosses represent the SPP peak positions calculated from the effective index of the SPP mode supported by the flat silver-PCDTBT:PCBM-air multilayer structure. (b, c) Electric field profiles for the surface plasmon modes in the 80 nm thick PCDTBT:PCBM on silver grating and (d, e) electric field profile for the surface plasmon modes in the 30 nm thick PCDTBT:PCBM on silver grating. To clarify the structure the bottom grating is the silver nanograting fabricated by nanoimprint lithography. The top surface is also corrugated simply by the result of depositing on a corrugated surface (From Ref. [62]).

When the grating height is increased, the surface plasmons lose their original characteristics of flat interfaces and the LSP modes become the dominant features. Kravets et al. studied a gold grating with 130 nm height fabricated by depositing gold on a 130 nm thick poly(methyl methacrylate) (PMMA) grating [63]. As shown in Figure 2, this structure exhibited extremely low reflectance and transmittance for TM polarized light, achieving blackbody-like absorption in the visible spectrum with a thickness much smaller than the wavelengths. A theoretical analysis is needed to understand this remarkable behavior. A rigorous theoretical treatment would require full consideration of two sets of gold gratings, one formed in between the PMMA lines and another formed on top of PMMA lines. However, since the grating lines are well separated (by more than 100 nm), near-field coupling between adjacent grating lines should be small and therefore it should be possible to use a much simpler effective medium theory. To generalize the Maxwell Garnett results for spherical particles given in equations (14)~(16) to non-spherical shapes, we introduce depolarization factor, L. For a general ellipsoidal shape, L is defined as [64]

$$L_i = \frac{a_x a_y a_z}{2}\int_0^{\infty} \frac{dS}{\left(s+a_1^2\right)\sqrt{\left(s+a_x^2\right)\left(s+a_y^2\right)\left(s+a_z^2\right)}}$$

$$i = x, y, z \tag{25}$$

where a_i is the semi-axis along the *i*-direction. For a sphere, $L_x = L_y = L_z = 1/3$. For a cylinder oriented along the x-direction, $L_x = 0$, $L_y = L_z = 1/2$. For a thin film in the yz plane, $L_x = 1$, $L_y = L_z = 0$. We can now write a generalized Maxwell Garnett effective permittivity as follows [65].

$$\varepsilon^{i}_{eff} = 1 + \frac{f\alpha_i}{1 - fL_i\alpha_i} \quad \text{where } i = x, y, z \tag{26}$$

where f is the volume fraction as before. Furthermore, the volume-normalized polarizability can be written in terms of the depolarization factor as,

$$\alpha_i = \frac{\varepsilon_s/\varepsilon_m - 1}{1 + L_i(\varepsilon_s/\varepsilon_m - 1)} \quad \text{where } i = x, y, z \tag{27}$$

where ε_s and ε_m are the permittivity of the ellipsoid and medium, respectively. Combining the two equations above yields

$$\alpha_z = \varepsilon_s/\varepsilon_m - 1 \;\; and \;\; \alpha_y = \alpha_z \; \frac{2(\varepsilon_s/\varepsilon_m - 1)}{\varepsilon_s/\varepsilon_m - 1} \tag{28}$$

$$\varepsilon^{x}_{eff} = 1 + f(\varepsilon_s/\varepsilon_m - 1) \;\; and$$
$$\varepsilon^{y}_{eff} = \varepsilon^{z}_{eff} \; \frac{\varepsilon_s/\varepsilon_m - 1 + f(\varepsilon_s/\varepsilon_m - 1)}{\varepsilon_s/\varepsilon_m + 1 - f(\varepsilon_s/\varepsilon_m - 1)} \tag{29}$$

for a cylinder oriented along the x-axis. Equation (29) yields the real and imaginary parts of the effective refractive index of the gold grating to be roughly 1.5 and 0.3, respectively, over a broad wavelength range up to 500 nm for TM polarization. These values are close to the optimal complex refractive index value for near-perfect absorption in a thin film. For example, a film with 130 nm thickness on a glass substrate would exhibit 90% absorption at $\lambda = 400$ nm if the refractive index is $1.3 + 0.5i$. A similar approach can be made with nanoparticles instead of nanograting [66]. In this case, the polarizability of a sphere given in equation (15) in Section 2.3 can be used in the Maxwell Garnett effective medium theory. The major difference in the nanoparticle composite medium and the nanograting is the polarization dependence. The inherent isotropy of spherical geometry naturally leads to isotropic, polarization-independent medium in contrast to the strong polarization sensitivity in nanogratings.

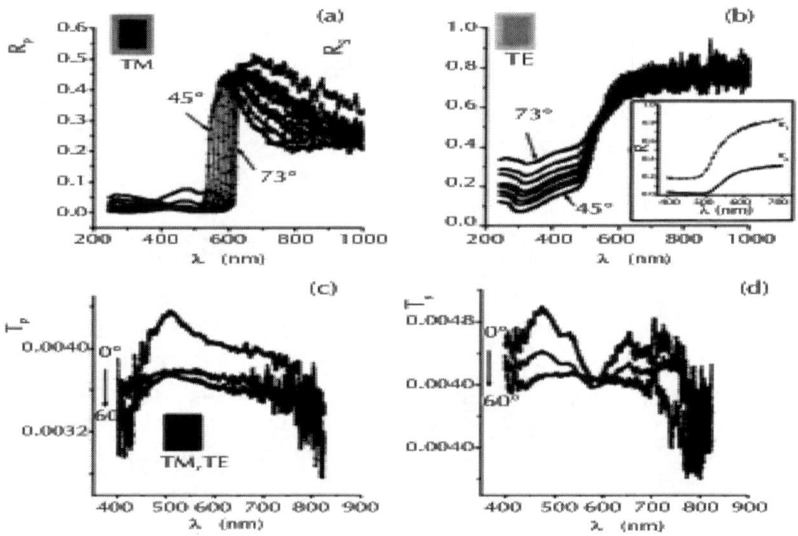

Figure 2: Reflectance and transmittance spectra of the gold nanostripe array with period = 320 nm and linewidth = 150 nm. Insets show the polarization-contrast optical micros- copy images (From Ref. [63]). Reflectance and transmittance spectra of the gold nanostripe array with period = 320 nm and linewidth = 150 nm. Insets show the polarization-contrast optical microscopy images. The inset in (b) shows reflectance spectra for s- and p-polarizations at normal incidence (From Ref. [63]).

A question of both scientific and technological interest is how much absorption is possible in ultrathin films. A simple solution may be found by using total internal reflection. Using the Kretschmann geometry, an ultrathin film absorber can be excited at glancing angle, effectively extending the optical path to infinity and consequently achieving near-perfect absorption. Using this method, 94% absorption in a 4.5 nm thick NbN nanowire array has been demonstrated [67]. It should be noted that this method does not require surface plasmon or any other types of surface resonances, although they could certainly help. While this approach clearly offers a quick solution to achieve perfect absorption, it is not always possible or feasible to use total internal reflection geometry and it would be much more desirable to achieve high absorption at normal incidence or over a broad range of incident angles. To tackle this problem, let us consider a thin film with thickness d and refractive index $m=n+i\kappa$ on a semi-infinite substrate with refractive index n_s and incident medium with refractive index n_i (Figure 3a). The reflection and transmission coefficients for normal incidence can then be written in a general form as follows.

$$r = \frac{r_i - r_s\gamma^2}{1 - r_i r_s \gamma^2} \quad and \quad t = \frac{(1+r_i)(1+r_s)\gamma}{1 - r_i r_s \gamma^2}\exp(-in_s\delta)$$
$$where \ r_{i,s} = \frac{n_{i,s} - m}{n_{i,s} + m} \quad and \ \gamma = \exp(im\delta) \tag{30}$$

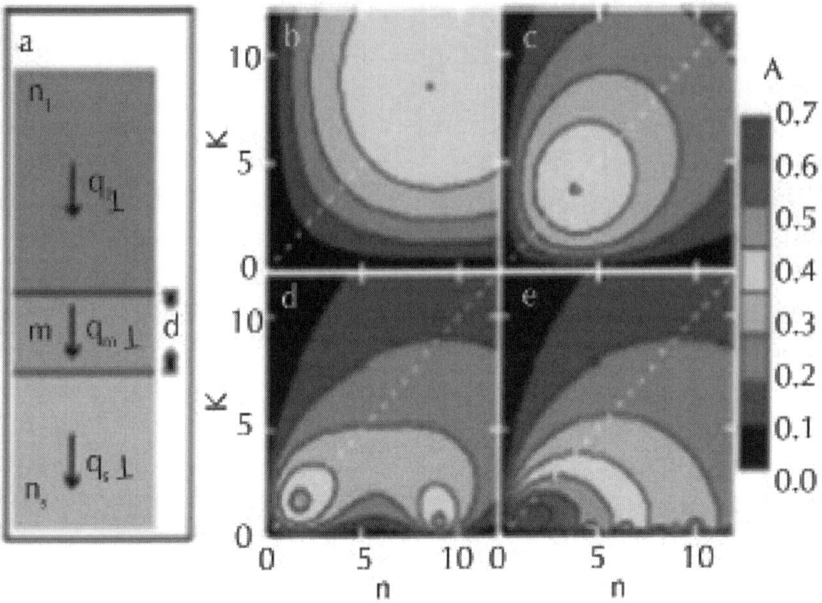

Figure 3: Absorptance of thin film with complex refractive index.(a)Thin homogeneous film with complex refractive index m=n+iκ and thickness d is embedded between two semi-infinite dielectrics of refractive indices n_i and n_s, respectively. In (b-e), absorptance in the film is shown as a function of its optical constants for normal incident light. The wavelength is 900 nm, and the dielectric environment has $n_i=n_s=1$. The film thicknesses d are 2, 10, 50, and 250 nm in panels b, c, d, and e, respectively. The dotted lines mark n=κ about which the shown absorption is nearly symmetric for the two thinnest films (b,c). In these cases, the absorptance reaches a peak value of close to 50% on the symmetry line. For the two thicker films (d,e), the peak height increases and additional maxima appear along the n-axis due to Fabry-Perot modes. (From Ref. [68]) Absorptance of thin film with complex refractive index.(a) Thin homogeneous film with complex refractive index m=n+iκ and thickness d is embedded between two semi-infinite dielectrics of refractive indices n_i and n_s, respectively. In (b-e), absorptance in the film is shown as a function of its optical constants for normal incident light. The wavelength is 900 nm, and the dielectric environment has $n_i=n_s=1$. The film thicknesses d are 2, 10, 50, and 250 nm in panels b, c, d, and e, respectively. The dotted lines mark n=κ about which the shown absorption is nearly symmetric for the two thinnest films (b, c). In these cases, the absorptance reaches a peak value of close to 50% on the symmetry line. For the two thicker films (d, e), the peak height increases and additional maxima appear along the n-axis due to Fabry-Perot modes (From Ref. [68]).

Also, $\delta = q_0 d$ and q_0 is the vacuum wave number. Generalization to non-normal incidence is straightforwardly done by replacing q_0 with the normal component of incident wave vector. The absorptance is given by where $C = n_s/n_i$ for transverse-electric (TE) incident field and for TM. One can now survey the possible values of m to find the maximum absorption. Figure 3(b ~ e) shows the absorptance at $\lambda_0 = 900$ nm for various film thicknesses of 2, 10, 50 and 150 nm. All cases show a maximum absorptance of approximately 50% along the line $n = \kappa$. Thick films show additional maxima along the horizontal axis due to the Fabry-Perot resonances. In ultrathin films, only those maxima occurring along $n = \kappa$line are achievable. Normally, for a fixed value of m, the absorption vanishes with vanishing thickness as $A = O(\delta)$. To go beyond this limit, Hägglund et al. sought a refractive index value with dependence where $-1 < v < 0$ so that m is large enough to produce substantial absorption while remains small [68]. The optimum value of m_{max} resulting maximum absorption was given as,

$$m_{\max} = \frac{1+i}{\sqrt{2}}\left(\frac{n_i + n_s}{\delta}\right)^{1/2} + \frac{1+i}{6\sqrt{2}}\frac{\left(n_i^3 + n_s^3\right)}{\left(n_i + n_s\right)^{3/2}}\delta^{1/2} + O\left(\delta^{3/2}\right) \tag{31}$$

The maximum absorptance is given as

$$A_{\max} = \frac{n_i}{n_i + n_s}\left[1 + \frac{n_s}{6}(3n_s - n_i)\delta^2\right] + O\left(\delta^4\right) \tag{32}$$

Equations (31) and (32) show that the maximum absorption occurs at $n = \kappa$ and the maximum absorptance is close to 50%, both of which are shown in Figure 3.

It is obvious that no natural materials would exhibit the optimum refractive index given in equation (31). It is therefore necessary to create a composite medium whose effective refractive index is close to the optimum. It was shown that 50% absorptance could be realized over a broad range of wavelengths using an array of ellipsoidal silver shells with various geometry [68]. This scheme can be extended to a thin film with a back reflector. The reflector automatically eliminates transmission and the problem is reduced to minimizing reflectance. A general system to consider is an ultrathin film with a complex refractive index *m* and thickness *d*, separated from a perfect reflector by a spacer layer with refractive index *ns* and thickness *h*. The incident medium index is n_i. The reflection coefficient is then given as

$$r = \frac{r_i - r_d \gamma^2}{1 - r_i r_d \gamma^2} \tag{33}$$

Where

$$r = \frac{n_{i,d} - m}{n_{i,d} + m}, \quad \gamma = \exp(im\delta)$$

and

$$n_s = in_s \ \cot(\pi \lambda_c / 2\lambda) \tag{34}$$

The critical wavelength λ_c is defined as the wavelength satisfying the quarter wavelength condition,$\lambda_c = 4n_s h$ Equation (33) has the same form as equation (30) and thus n˜smay be treated as the effective index of the spacer-reflector system. The maximum absorption condition is then found by requiring $r_i = r_d \gamma^2$ in equation (33). Retaining only the leading term the optimum refractive index is given as [69],

$$m^2 = \frac{i(n_i - .n_s)}{\delta} = \frac{\lambda}{2\pi d}\left[in_i + n_s \cot\left(\frac{\pi\lambda_c}{2\lambda}\right)\right] \quad (35)$$

When $\lambda=\lambda_c$, $\tilde{n}_s=0$ and the optimum refractive index is given as $m=(1+i)\sqrt{n_i / s\delta}$. This represents the critical coupling condition where the required refractive index *m* is minimum. On the other hand, when the spacer layer thickness tends to zero, n*N* diverges and the optimum refractive index cannot be satisfied. Again, the implementation of this concept requires an artificial composite medium made of plasmonic nanostructures. Recently, strong absorption has been experimentally observed using a gold nanoparticle array fabricated by block copolymer lithography and overcoated with a dielectric layer by atomic layer deposition [70]. As shown in Figure 4, the ultrathin composite film with a thickness of ~25 nm exhibited near-perfect absorption at around 600 nm. Vast majority of absorption occurs in the gold nanoparticles, which is not desirable for photovoltaic applications where the absorption must take place in the semiconducting material for current extraction. It is nevertheless noteworthy that the effective absorption coefficient of the gold nanoparticles in this structure was an order of magnitude greater than solid metal. A similar critical coupling leading to perfect absorption has been reported on a gold nanodisk array deposited on a glass substrate [71]. In this case, the critical coupling condition was achieved by tuning the nanodisk density and incident angle of incoming light. When the critical coupling condition is satisfied, reflected light is annihilated by the destructive interference between the light reflected from the substrate and the gold nanodisk array layer.

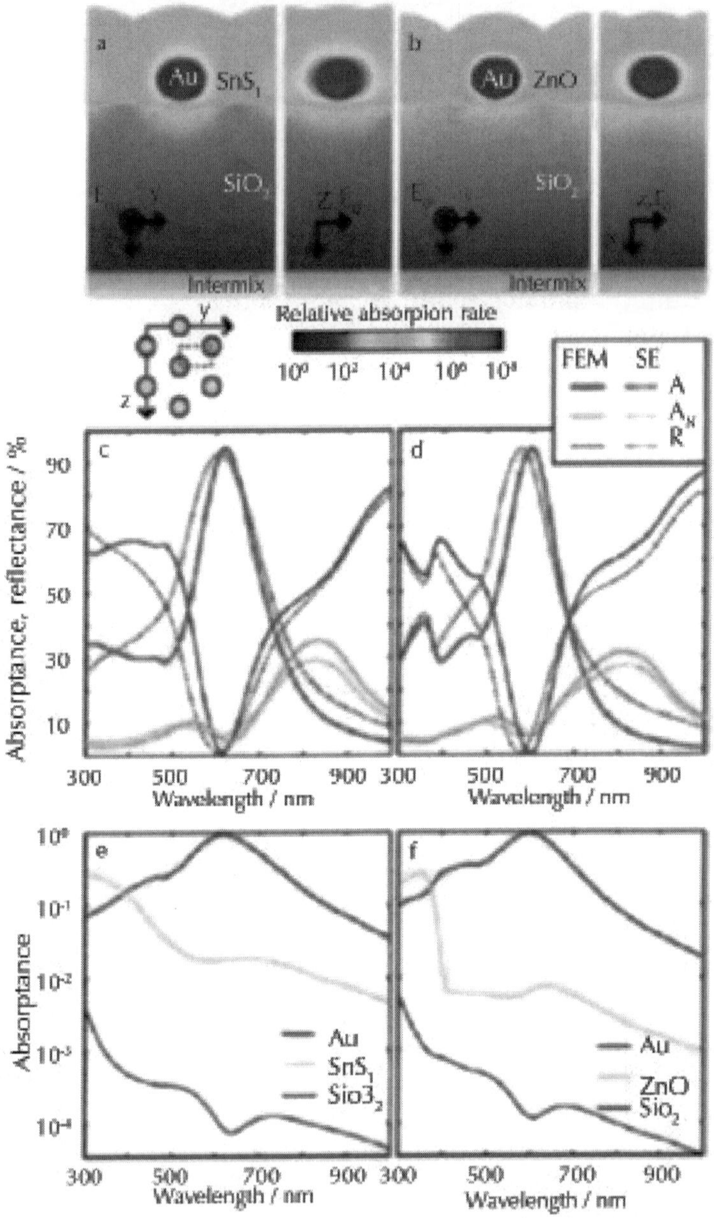

Figure 4: Absorptance of gold nanoparticle embedded thin films.(a)Finite element method (FEM) calculated absorption rates, determined for the sample with the highest absorbing SnS_x coated Au array at its peak wavelength. A plane wave with incident electric field strength E_0 propagates along the normal (x-) direction, with its electric field vector along z. The rectangular unit

cell used for the hexagonal lattice is shown below the plot (dashed lines). (b) Analogous for ZnO coated sample. (c,d) Sample reflectance (R), array absorptance (A), and absorptance in the Al reflector (AAl, including intermix layer), as calculated by FEM and obtained from the spectroscopic ellipsometry (SE) model. (e) FEM calculated absorptance in the Au dots of the SnS_x coated sample, in the SnS_xcoating, and in the SiO_2 spacer, respectively. (f) Analogous for the ZnO coated sample. (From Ref. [70]) Absorptance of gold nanoparticle embedded thin films.(a) Finite element method (FEM) calculated absorption rates, determined for the sample with the highest absorbing SnS_xcoated Au array at its peak wavelength. A plane wave with incident electric field strength E_0propagates along the normal (x-) direction, with its electric field vector along z. The rectangular unit cell used for the hexagonal lattice is shown below the plot (dashed lines). (b)Analogous for ZnO coated sample. (c, d) Sample reflectance (R), array absorptance (A), and absorptance in the Al reflector (AAl, including intermix layer), as calculated by FEM and obtained from the spectroscopic ellipsometry (SE) model. (e) FEM calculated absorptance in the Au dots of the SnS_x coated sample, in the SnS_x coating, and in the SiO_2 spacer, respectively. (f) Analogous for the ZnO coated sample (From Ref. [70]).

Invisibility

In addition to enhancing absorption, controlling or suppressing light scattering is also of great interest for many applications. It has been shown that the dominant dipolar contribution to the scattering from a spherical or cylindrical dielectric object could be greatly reduced when coated by a layer of material with negative or very low values of the real part of permittivity [72]. This technique is inherently non-resonant and thus the bandwidth of the scattering cancellation can be quite broad and limited only by the material dispersion of the coating. It was shown that the operational bandwidth could be further extended by use of multiple layers of coatings [73]. Potential applications of this type of cloaking include cloaked sensors [74], cloaked near-field probes [75,76] and reduction of optical forces [77]. For implementation, a natural choice for the coating material is a metal, which exhibits negative permittivity values at frequencies below its plasma frequency. However, in the optical frequency region, most metals tend to exhibit large negative permittivity values. Consequently, an unrealistically thin layer is required to cancel the dipole moment of a dielectric object whose permittivity is typically a small positive number. To circumvent this issue, an effective medium formed by plasmonic nanostructure is needed. Recently, Tamma et al. reported scattering cancellation by a gold nanograting fabricated on the sidewall of a silicon nanorod [78]. As shown in Figure 5(a), the cloaked object was a silicon nanorod fabricated by electron-beam lithography. The gold nanograting was then fabricated on the sidewall by a combination of gold evaporation, focused ion beam milling and silica protective coating deposition by sputtering. The final structure was coated with SU-8 to form a waveguide layer and the structure was excited by 1550 nm light coupled into the SU-8 guiding layer. The scattering properties were then investigated by near-field scanning optical microscopy (NSOM). The waveguide geometry used in this study makes it impossible to directly measure the scattered light intensity. Instead, NSOM was used to map the interference fringes produced by the incident light and back-scattered light (Figure 5b). It was clear that the cloaked silicon nanorod scattered much less because the interference fringes were weak and required a long integration time and high gain to obtain reliable data whereas the reference nanorod showed a strong interference pattern. Tamma et al. then used the effective medium theory to relate the effective index of the coating with the fringe curvature. Here the gold nanograting coating was described as a multilayer structure for which the effective medium theory simple yields [79]

$$\varepsilon_{\|} = f_a \varepsilon_a + f_b \varepsilon_b \quad and \quad \varepsilon_{\perp} = (f_a/\varepsilon_a + f_b/\varepsilon_b)^{-1} \tag{36}$$

where the subscripts || and ⊠ indicates polarization parallel and perpendicular to the interface, respectively. The result of the complete analysis shown in Figure 5(c) indicates that a 9.8 dB reduction in scattering cross section has been achieved. The same approach has been implemented with silver nanoparticles attached on silica spheres [80]. The scattering reduction was however small, ~20%, and the system also exhibited strong absorption.

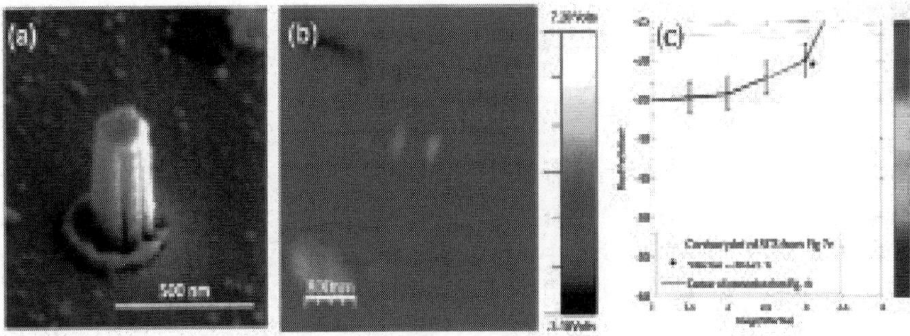

Figure 5: Scattering cancellation by plasmonic nanostructure.(a)Scanning electron micrograph (SEM) of the silicon nanorod with gold nanograting fabricated on the side wall. (b) Near-field scanning optical micrograph (NSOM) of scattered light by the cloaked silicon nanorod measured with λ=1550 nm. The fringe curvature contains the information on scattering cross section. (c) Scattering cross section extracted from the analysis of fringe curvatures. A reduction of 9.8 dB was achieved (From Ref. [78]) Scattering cancellation by plasmonic nanostructure.(a) Scanning electron micrograph (SEM) of the silicon nanorod with gold nanograting fabricated on the side wall. (b) Near-field scanning optical micrograph (NSOM) of scattered light by the cloaked silicon nanorod measured with λ=1550 nm. The fringe curvature contains the information on scattering cross section. (c) Scattering cross section extracted from the analysis of fringe curvatures. A reduction of 9.8 dB was achieved (From Ref. [78]).

In concluding the discussion on weakly coupled system that can be described well by the effective medium theory, it is worth noting the robustness of effective medium theory. As discussed in the previous section, the effective medium theory is valid only when the feature size is small compared to the wavelength and the interaction among the nanostructures is weak. When these two conditions are met, the optical properties of the composite structure are determined by the dipole moment of the individual nanostructure without having to include higher order resonances and any higher order interaction. These conditions certainly sound very restrictive but in reality the effective medium theory has been successfully applied to a wide variety of systems which at first glace do not seem to satisfy the conditions for validity. An excellent example is the silicon nanorod cloaked by an array of gold nanowires described above. The individual gold nanowire was 13 nm thick and 20 nm wide and the spacing between two adjacent gold nanowires was 63 nm. This system was modeled as a multilayer composed of 20 nm thick gold and 43 nm thick silica films. Despite this seemingly gross simplification, the effective medium theory provided a remarkably accurate result in good agreement with experiments. It can be argued that the effective medium theory, although oversimplified, has captured the essential part of the light-matter interaction, which in this case concerns the induced surface charge and resultant screening effect. The effective medium theory for multilayers resulting in effective permittivity given in equation [35] considers the induced polarization in each layer. When the incident electric field is perpendicular to the multilayer interfaces, it produces a net interface charge due to the discontinuity in permittivity at the interface. These interface charges then produce a polarization field that counteracts the incident electric field, resulting in a permittivity given as $\varepsilon_{\boxtimes}$ in equation [35]. On the other hand, when the incident electric field is parallel to the interface, no interface charge is formed and thus the effective permittivity is simple given as an arithmetic mean of the constituent permittivities, as given for $\varepsilon_{||}$ in equation [35]. The experiments in the cloaked silicon nanorod were done for TM waveguide mode which has electric field parallel to the gold nanowires. This corresponds to the no interface charge case and the multilayer effective medium theory for no screening case is applicable. If the experiments were done for perpendicular polarization, it would have been important to accurately account for the interface charge density and thus the multilayer effective medium theory could have been inaccurate.

Another case worth mentioning is the nanoparticle array. It is well established that small nanoparticles have dominantly dipole character with negligible contributions by the high order components. Thus, as long as the nanoparticles are separated well enough, effective medium theory should work well. But defining a precise validity limit is difficult because unlike the perturbation theory the effective medium theory does not allow quantitative estimation of errors. For plasmonic nanoparticles, the literature suggests volume fraction of 10~20% would be the upper limit where the effective medium theory begins to break down [49,81]. In a study of collective magnetic resonance supported by a cluster of plasmonic nanoparticles, Tamma et al. presented a comparison between the rigorous multiple scattering theory and the extended Maxwell Garnett effective medium theory [49]. Figure 6 shows the dispersion curves and effective permittivity for silver nanoparticle arrays with silver volume fractions of 0.1, 0.3 and 0.5. As shown, even at a very high volume fraction of 0.5, the effective medium theory agrees reasonably well with the rigorous multiple scattering theory. The discrepancy is found near the band edge in all cases where the discrepancy remains small for f=0.1 and becomes significant for higher volume fractions. This discrepancy observed near the band edge captures the nature of approximation in the effective medium theory. Unlike the effective medium theory, the multiple scattering formalism requires the complete knowledge of the nanoparticle positions in the array. For the calculations in Figure 6, the nanoparticles were assumed to be in the face-centered cubic (fcc) structure and the light propagation along the (111) direction was calculated. If the calculations were made for other crystallographic directions, the resulting dispersion curves would be different and the differences would be most pronounced near the band edge where the Bragg resonance condition is met. By construction, the effective medium theory does not require the knowledge of nanoparticle arrangement. This implies the effective medium theory would represent an averaged response of all different crystallographic orientations. It is therefore not surprising to see the major difference between multiple scattering theory and effective medium theory near the band edge. This however does not necessarily represent the deficiency of effective medium theory because most nanocomposite structures do not have a definite crystal structure. A few exceptions are DNA-driven self-assembly of plasmonic nanoparticles that lead to single crystalline fcc and body-centered cubic structures [82,83]. But even in those cases, the achievable structures remain small and multiple crystallographic orientations may be in play. Therefore, although it is advisable to compare with rigorous multiple scattering or other methods whenever possible, the effective medium theory remains highly useful in an extremely wide range of plasmonic nanostructures.

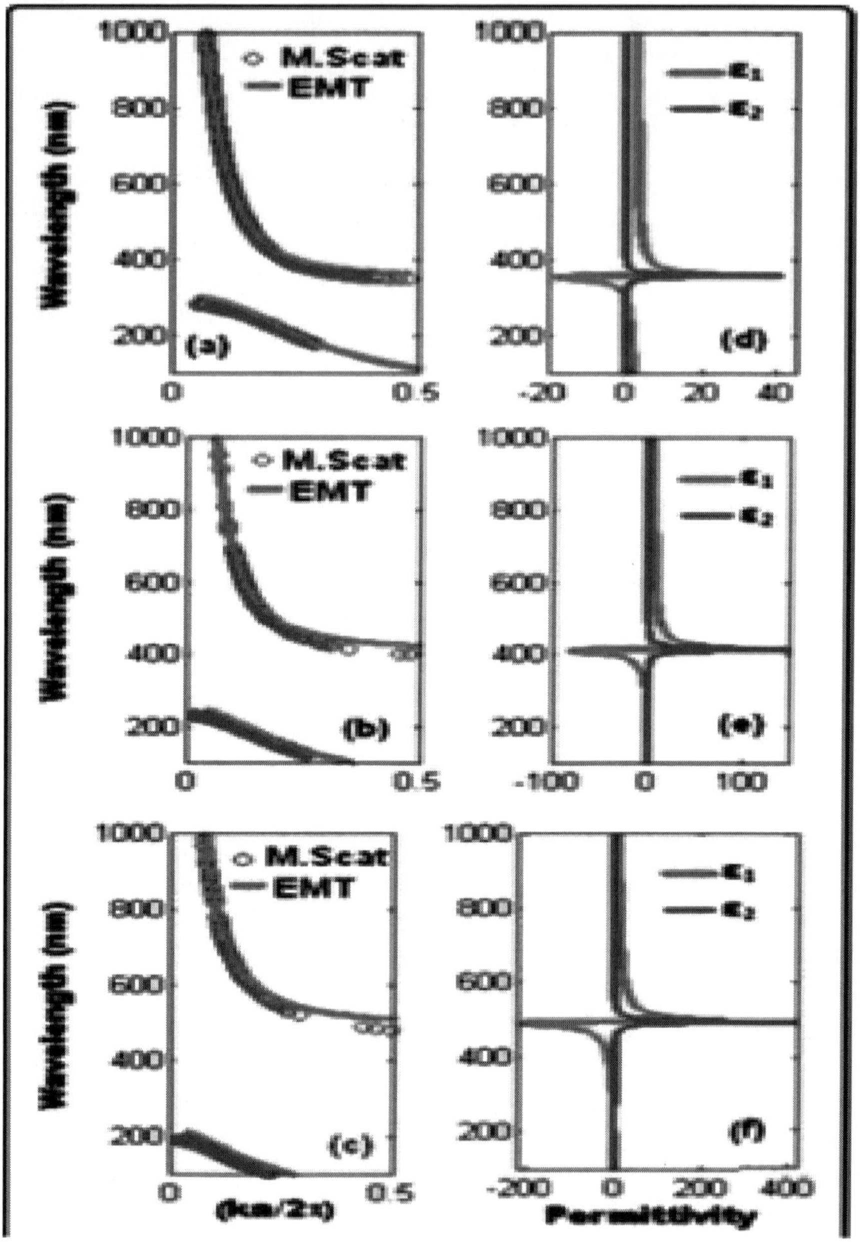

Figure 6: Comparison of multiple scattering theory and effective medium theory.(a), (b), (c) Dispersion curves obtained from multiple scattering (M.Scat) and extended Maxwell-Garnett effective medium theory (EMT) for silver nanoparticle array with volume fractions of 0.1, 0.3 and 0.5, respectively. (d), (e), (f) Real and imaginary parts of the effective permittivity calculated by

the extended Maxwell-Garnett effective medium for silver volume fractions of 0.1, 0.3 and 0.5, respectively (From Ref.[49]). Comparison of multiple scattering theory and effective medium theory.(a), (b), (c)Dispersion curves obtained from multiple scattering (M.Scat) and extended Maxwell-Garnett effective medium theory (EMT) for silver nanoparticle array with volume fractions of 0.1, 0.3 and 0.5, respectively. (d), (e), (f) Real and imaginary parts of the effective permittivity calculated by the extended Maxwell-Garnett effective medium for silver volume fractions of 0.1, 0.3 and 0.5, respectively (From Ref. [49]).

Strongly Coupled Plasmonic Nanostructures

Dimers

Strongly coupled plasmonic nanostructures have been studied extensively during the past decade. For example, the hot spot leading to enhanced Raman scattering is generally attributed to the coupling between two adjacent silver nanoparticles producing strong local field in the gap between the two nanoparticles [84]. A similar phenomenon has been observed in a coupled-dipole nano-antenna composed of two nanorods aligned end-to-end and separated by a small gap [85]. Also, early experimental studies on the simple model system of two closely spaced metal nanoparticles, a system often dubbed as dimer, showed the plasmon resonance exhibits red shift with decreasing gap between the nanoparticles [86-89]. Both the formation of hot spots and the red shift with decreasing gap size are due to the hybridization of individual nanoparticle plasmons. When the gap between the nanoparticles is small enough to allow significant overlap of the plasmon modes of the individual nanoparticles, the strong near-field interaction consequently leads to hybridization of the individual single particle plasmon modes to form a new set of coupled modes delocalized over the entire structure. In a dimer, the single particle plasmon modes couple with each other to form symmetric and anti-symmetric combinations. This is the standard coupled oscillator problem with numerous analogous examples in physics [90]. In the plasmonic dimer, the oscillating entity is the free electron charge in the metal. When the charge density in the two particles oscillates in phase along the axis of the dimer, we have a symmetric mode with strong electric field in the gap. When the two oscillations are 180° out of phase, we have an anti-symmetric mode whose electric field profile has a null in the gap. In molecular analogy, the symmetric mode corresponds to the bonding orbital whereas the anti-symmetric mode is the anti-bonding orbital. Also, the symmetric mode has a lower energy and the anti-symmetric mode has a higher energy, just as the bonding orbital has a lower energy and the anti-bonding orbital has a higher energy. As illustrated in Figure 7 for a nanorod dimer, this hybridization model offers a good explanation for both the hot spot formation and the red shift in plasmon resonance wavelength.

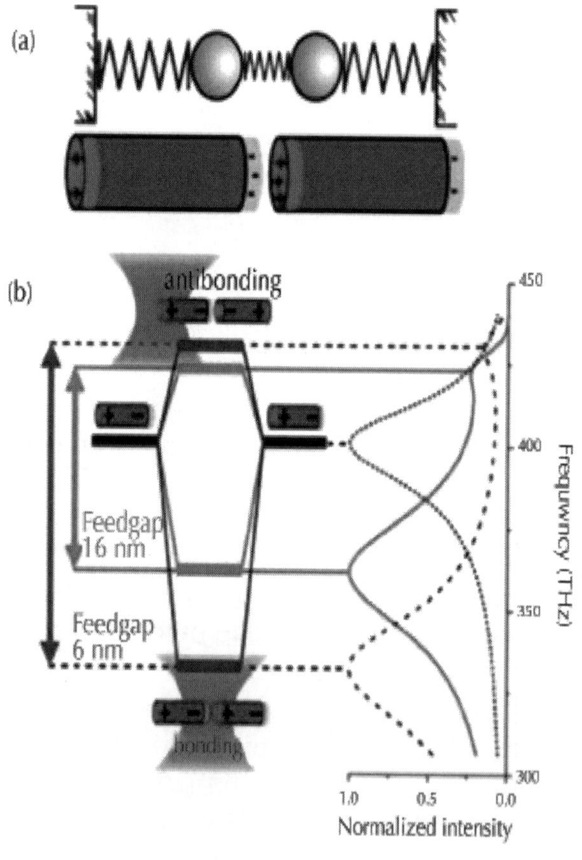

Figure 7: Plasmon coupling in nanorod dimer.(a) Schematic drawings of a coupled-dipole nanoantenna made of two plasmonic nanorods with a small gap and its mechanical analog, a coupled mechanical oscillator. (b) Energy-

level diagram and simulated near-field intensity spectra for 30 nm high, 50 nm wide and 110 nm long symmetric two-nanorod antennas with 6 nm (blue dashed) and 16 nm (green solid) gap, as well as single-nanorod antennas with the same dimensions (black dotted) (From Ref. [90]). Plasmon coupling in nanorod dimer.(a) Schematic drawings of a coupled-dipole nanoantenna made of two plasmonic nanorods with a small gap and its mechanical analog, a coupled mechanical oscillator. (b) Energy-level diagram and simulated near-field intensity spectra for 30 nm high, 50 nm wide and 110 nm long symmetric two-nanorod antennas with 6 nm (blue dashed) and 16 nm (green solid) gap, as well as single-nanorod antennas with the same dimensions (black dotted) (From Ref. [90]).

The hybridized plasmon modes in a dimer can be monitored by directly solving Maxwell's equations numerically using techniques such as finite-difference time-domain (FDTD) [91] or finite-element method (FEM) [92]. Nevertheless, it is always profitable to seek for a theoretical model that could provide deeper insights on the physics governing the hybridization. Also, it is worth mentioning the inability of most brute force numerical simulations to excite anti-symmetric modes. This deficiency stems from the typical simulation geometry in which the nanostructure under investigation is excited by incident light. The simulation then calculates the resultant absorption or scattering which reveals the characteristics of the hybridized plasmon modes. This unfortunately works well only for the symmetric mode, which typically possesses a large dipole moment and thus interacts strongly with light. The anti-symmetric mode, however, represents a state in which the induced dipole moments in the individual nanoparticles align anti-parallel and thus only produce a small net dipole moment. This mode therefore does not interact with light and remain invisible in most experiments and simulations. For this reason, the symmetric mode is often referred to as the bright mode and the anti-symmetric mode as the dark mode. The anti-symmetric mode is nonetheless important as, for example, it can interact with the bright mode to produce Fano resonance as described later. It is therefore useful to have a theoretical model that can describe both the symmetric and anti-symmetric modes equally well. The simplest model employs the electrostatic approximation where the retardation effects are neglected. In this formalism, the polarizability, α, of individual nanoparticles is given by equations (15) and (27) for spheres and ellipsoids, respectively. The induced dipole moment in each particle is then given as $\boldsymbol{\mu} = \alpha\mathbf{E}$.

In a dimer, the field, **E**, contains both the applied field and the field produced by the neighboring particle. Since an individual nanoparticle is treated as a point dipole, we can write

$$\mathbf{E}_1 = \mathbf{E}_{inc} + g \cdot \frac{\boldsymbol{\mu}_2}{\varepsilon_m d^3} \text{ and } \mathbf{E}_2 = \mathbf{E}_{inc} + g \cdot \frac{\boldsymbol{\mu}_1}{\varepsilon_m d^3} \tag{37}$$

Here the subscripts 1 and 2 index the two particles in the dimer, ε_m is the permittivity of the background medium and g is the geometry dependent parameter given as g=2 for end-to-end coupling and g=-1 for side-to-side coupling [93,94]. The coupled equation in [36] can be solved to find the polarizability of the bonding mode,

$$\langle \alpha \rangle = \frac{1}{2} \cdot \frac{\alpha_1 (1 + g\alpha_2/d^3) + \alpha_2 (1 + g\alpha_1/d^3)}{1 - g^2 \alpha_1 \alpha_2 / d^6} \tag{38}$$

The plasmon resonance condition is found by the maximum polarizability and, if the two particles are identical, is given as,

$$\alpha = \frac{d^3}{g} \tag{39}$$

Using the polarizability given in equation (27), we find

$$\frac{\varepsilon_s}{\varepsilon_s} = \frac{1 + d^3 (1 - L_i)/g}{1 + d^3/g} \tag{40}$$

Since the permittivity of gold has almost linear dependence on wavelength in the visible region, equation (39) specifies the dependence of the surface plasmon wavelength of a gold dimer on the center-to-center spacing. This simple model correctly predicts the red shift of the symmetric mode for end-to-end aligned dipole pair and the blue shift for side-to-side aligned dipole pair. The opposite behavior originates from the opposite sign of g in equations (37) ~ (39). In plasmonic dimer made of two identical particles, the end-to-end dipole alignment corresponds to the excitation of dimer with incident polarization along the dimer axis while the side-to-side coupling case corresponds to incident polarization perpendicular to the dimer axis. The opposite peak shifts in the two cases have been experimentally observed in the extinction spectra of gold nanodisk pairs fabricated by electron-beam lithography (Figure 8a) [88]. Furthermore, Jain et al. reported the fractional plasmon peak shift, $\Delta\lambda/\lambda_o$, exhibits a universal scaling against the normalized gap width defined as gap/diameter, irrespective of the detailed geometrical parameters, as shown in Figure 8(b) [95]. This universal scaling was also found for dimers of non-circular shapes and nanoshells and also in trimers [96,97]. The origin of the universal scaling behavior may be found in equation (39), which predicts that the fractional peak shift would behave as $(d/D)^{-3}$ or $[(s+D)/D]^{-3}$ where $s=d-D$ is the gap width between the particles and D is the diameter. When the gap width becomes extremely small, the simple dipole-dipole interaction model breaks down and one must take into account the higher order multipole terms, leading to an exponential dependence observed in Figure 8. This omission of multipole interaction terms is one of the key limitations in the dipole-dipole interaction model, which leads to a significant underestimation of plasmon coupling at extremely small gaps. Additionally, the dipole model is essentially based on the electrostatic approximation which ignores any retardation effect. To properly describe the interaction at small gaps including the retardation effects would require the use of numerical modeling techniques such as FDTD and FEM. Before moving on, it is worth mentioning that the universal exponential dependence has led to the derivation of the so-called plasmonic ruler equation, which enables precise determination of distances in biological systems, opening many opportunities for biosensing [95].

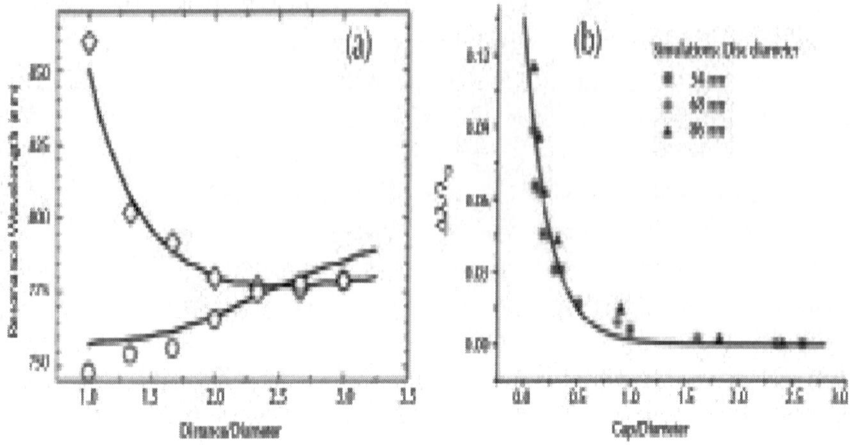

Figure 8: Plasmon coupling in gold nanoparticle dimer.(a) Spectral position of the extinction maximum vs. interparticle distance for seven gold nanodisk pairs fabricated by electron-beam lithography. (From [88]) (b) Calculated fractional plasmon peak shift vs the ratio of interparticle gap to nanodisk diameter, showing that the scaled data points for the different disc sizes follow a common trend, which can be fit together to the single-exponential decay (solid curves) $y=ae^{-x/\tau}$ with and $\tau=0.23\pm0.03$ (From Ref.[95]). Plasmon coupling in gold nanoparticle dimer.(a) Spectral position of the extinction maximum vs. interparticle distance for seven gold nanodisk pairs fabricated by electron-beam lithography (From [88]). (b) Calculated fractional plasmon peak shift vs the ratio of interparticle gap to nanodisk diameter, showing that the scaled data points for the different disc sizes follow a common trend, which can be fit together to the single-exponential decay (solid curves) $y=ae^{-x/\tau}$ with and $\tau=0.23\pm0.03$ (From Ref. [95]).

Note that the previous discussion of dipole-dipole coupling focused only on the symmetric mode where the two individual dipoles are parallel to each other. As stated earlier, the anti-symmetric mode with anti-parallel dipoles has a small net dipole moment and thus interacts only weakly with light. This is not true when the anti-symmetric mode possesses a large magnetic moment. In this case, the anti-symmetric mode interacts strongly with light through its magnetic field. A dimer of small spherical nanoparticles does not support strong magnetic resonance but a pair of nanorods can. A pair of gold nanorods was shown to support a strong magnetic resonances [98,99] and soon after used to construct negative index material [100]. A similar behavior was also observed in a pair of nanoholes in two closely spaced metal films, which is expected to exhibit the same properties as the nanorod pair through the Babinet principle [101]. The nanohole pair design was then adapted to the fishnet structure which is known to be the most effective design for negative index [102]. Later, nanorod pairs with various different arrangements were investigated [103]. The hybridization scheme for various geometries is shown in Figure 9 together with the surface change densities for various plasmon modes. Here the hybridization scheme was obtained by the plasmon hybridization model and the surface charge density by the boundary integral method, both of which are described in Section 2.4. As shown in Figure 9(a), the single nanorod supports two resonances, longitudinal and transverse modes where the transverse mode has a higher energy due to the smaller transverse dimension. When the two nanorods are placed close together, the near-field interaction leads to hybridization and mode splitting where the symmetric and anti-symmetric combinations form bonding and anti-bonding modes. In the perfectly symmetric arrangements of end-to-end and side-to-side pairing, the anti-symmetric combination results in perfect cancellation of dipole moments. Thus the resultant hybridized mode is dark and does not respond to the incident light. Note that, unlike the nanorod pair used to observe magnetic resonance, which has one nanorod on top of the other so that the induced magnetic moment can align with the incident magnetic field, these nanorod pairs are all fabricated on the same plane and thus no magnetic interaction is expected. The energies of the hybridized modes are easily understood by the electrostatic interaction between electronic charge densities. For example, in the end-to-end pair, the anti-symmetric mode has the same charge concentrated near the gap, resulting in high energy while the symmetric mode has the opposite charge across the gap, leading to lower energy. Thus in this case, the symmetric mode becomes the bonding mode and the anti-symmetric mode forms the anti-bonding mode. For the side-to-side pair, the opposite is the case and the symmetric mode has higher energy than the anti-symmetric mode, making the anti-symmetric mode the bonding mode and the symmetric mode the anti-bonding mode. For other non-symmetric arrangements, the hybridized mode energy depends on the exact geometry. It should be noted that both the plasmon hybridization model and the boundary integral method used to generate Figure 9 take into account only the electrostatic interaction. The energy levels in Figure 9(a) and (b) are thus different from the experimentally measured plasmon peak positions, generally blue shifted from the experimental values due to the neglect of retardation effect.

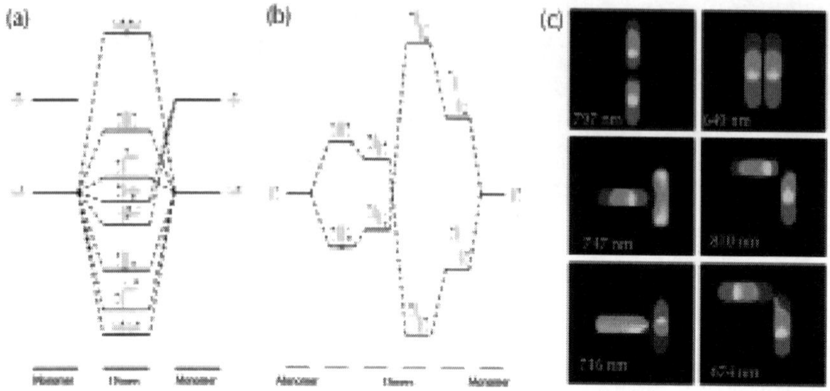

Figure 9: Plasmon hybridization schemes for (a) nanorod dimers in different geometric arrangements and (b) nanorods initially arranged side-to-side and then increasingly longitudinally offset as a function of the center-to-center offset. (c) Surface charge density of interacting gold nanorods placed 1.5 nm apart and with dimensions 78 nm × 24 nm, calculated using the boundary integral method (From Ref. [103]). Plasmon hybridization schemes for (a) nanorod dimers in different geometric arrangements and (b) nanorods initially arranged side-to-side and then increasingly longitudinally offset as a function of the center-to-center offset. (c) Surface charge density of interacting gold nanorods placed 1.5 nm apart and with dimensions 78 nm × 24 nm, calculated using the boundary integral method (From Ref. [103]).

Heptamers

To go beyond the simple dimer structure to more complex nanostructures, one needs a systematic way to classify various eigenmodes with complex symmetry properties. The group representation theory provides a power means to accomplish that [23]. For a given eigenvalue equation, $\mathbf{A}\mathbf{x}_n = a_n\mathbf{x}_n$, we can define a group of symmetry operators that leaves the operator **A** invariant. Note that the discussion here is generally applicable to any types of eigenvalue equations which could be the Schrödinger's equation in quantum mechanics, Maxwell's equation in optics and electromagnetics, or the fluid dynamics and surface charge density equations used in the plasmon hybridization model and boundary integral formalism, respectively, discussed in Section 2.4. If a symmetry operator, **S**, leaves the operator A invariant, it is straightforward to show that the two operators commute with each other and share the same eigenstates. If all symmetry operators that leave the operator **A** invariant are found, then it is possible to classify all eigenstates of operator **A** in terms of how they transform under the various symmetry operators. In the language of group theory, all eigenstates of operator **A** can be indexed by the irreducible representations of the symmetry group. Now let us proceed to the plasmon modes in a trimer using the group representation theory. For a trimer composed of three identical nanoparticles placed at the corners of an equilateral triangle, the symmetry operators that leave the system invariant form the point group *D3h* and thus the eigenmodes of a trimer can be classified by the irreducible representations of *D3h*. For simplicity, we treat the plasmon mode of individual nanoparticle as a dipole mode, which is a good approximation for small particles. Then, the individual plasmon mode belongs to the $D^{(l=1)}$ irreducible representation of the full rotation group. For a trimer, we should consider the triple direct product of $D^{(l=1)}$ irreducible representations and their subsequent reduction into a combination of irreducible representations of point group *D3h*, which yields [104],

$$\Gamma_{trimer} = A_1^{'} + A_2^{'} + 2E' + A_2^{''}E'' \tag{41}$$

The transformation properties of the various irreducible representations of point group *D3h* can be found in the character table available in the literature [23]. From the character table, one can then construct the projection operators and generate the symmetry-adapted basis functions, as shown in Figure 10, which transform according to the symmetry properties of the irreducible representations. The one-dimensional irreducible representations, A'_1, A'_2, A''_1, A''_2, have only one basis function each and the two-dimensional irreducible representations, E', E'', have two which are degenerate. Figure 10 is a powerful illustration of the symmetry properties of the hybridized plasmon modes. First, we notice the planar geometry of trimer naturally separates the in-plane and out-of-plane modes, which belong to the primed and double-primed irreducible representations, respectively. For light incident normally to the plane of the trimer, the out-of-plane modes are orthogonal to the incident electric field oscillation. Therefore, only the in-plane modes would interact with the incident light and the out-of-plane modes would not respond. Furthermore, the two in-plane modes, A'_1, A'_2, represent perfectly symmetric combinations of the individual dipoles resulting in zero net dipole moment. These modes are therefore dark modes which do not interact with the incident light. The two in-plane *E′* modes possess net dipole moments and are the ones that will absorb or scatter light. It should be reminded that the actual plasmon modes excited by an incident light are dependent on the precise excitation conditions and would in general not be the *E′* modes shown in Figure 10. Rather, they may be expressed as a linear combination of the *E′* modes. A similar analysis has also been carried out for a nanorod trimer [17].

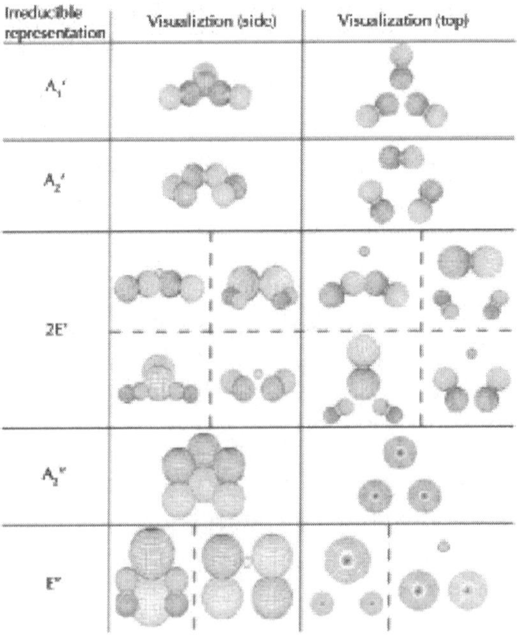

Figure 10: Irreducible representations and symmetric linear combinations of the$l$ =1 dipole modes as predicted by group theory and plasmon hybridization model (From Ref. [[104]]). Irreducible representations and symmetric linear combinations of the$l$ =1 dipole modes as predicted by group theory and plasmon hybridization model (From Ref. [[104]]).

The group theory provides a powerful and effective way to describe more complex structures and extensive studies, both theoretical and experimental, have been conducted on quadrumers [56],[104-106], pentamers [20,107], and hexamers [17,18,20]. The structure that received by far the most attention was the heptamer which is constructed by adding a central particle to a hexamer. The main reason for the interest has been the Fano resonance the heptamer supports. In most systems exhibiting resonance, the lineshape function describing the resonance is typically Lorentzian. However, when a discrete state interacts strongly with a continuum, a new resonance characterized by a distinctly asymmetric lineshape function may arise, as first discovered by Fano[108]. Since the origin of Fano resonance is the interference between the discrete and continuum quantum states, it is fundamentally a quantum phenomenon. However, the interference phenomenon is commonly observed in classical optics and Fano-like resonances have been observed in many systems. The oldest example is the Wood's anomaly arising from the interference between the tangentially diffracted wave and the incident wave [109]. It has also been observed in coupled waveguide-cavity systems, photonic crystals and plasmonic nanostructures for which many reviews are available [110-112]. In addition to the rich physics it reveals, Fano resonance also has many technological applications thanks to its high sensitivity to the environmental parameters [19]. Naturally, this led to the demonstration of optical sensing based on Fano resonances [113,114]. Here we review Fano resonances in heptamers in detail.

A heptamer is composed of seven nanoparticles in a geometry with 6-fold rotational symmetry. Six peripheral nanoparticles form a hexamer where the nanoparticles are placed at the corners of a regular hexagon. One central nanoparticle is located at the center of the hexagon. The heptamer has the same point group symmetry as hexamer, D_{6h}. Since the central nanoparticle is not connected to any other nanoparticles by the symmetry operators of point group *D6h*, the irreducible representations for a heptamer can be found by the union of those for the hexamer formed by the peripheral nanoparticles and the central nanoparticle, $\Gamma_{hep}=\Gamma_{hex}+\Gamma_{cent}$. Similarly to the trimer case, considering only the dipole modes of the individual nanoparticles, the irreducible representations of the hexamer, Γ_{hex}, are found by taking the direct product of six $l=1$ irreducible representations, $\Gamma^{(l=1)}$, of the full rotation group and reducing it into the irreducible representations of the D_{6h} symmetry group. Likewise, the irreducible representation of the central nanoparticle is found by simply reducing $\Gamma^{(l=1)}$. The results are as follows

$$\begin{aligned}\Gamma_{hex} &= A_{1g} + A_{2g} + B_{1u} + B_{2u} + 2E_{2g} + 2E_{1u}\\ \Gamma_{cent} &= E_{1u}\\ \Gamma_{hep} &= A_{1g} + A_{2g} + B_{1u} + B_{2u} + 2E_{2g} + 3E_{1u}\end{aligned} \quad (42)$$

where we only considered the in-plane modes in these reduction schemes. As explained earlier, the out-of-plane dipoles will not interact with the normally incident light. The symmetry-adapted basis functions for these irreducible representations are shown in Figure 11(a) [115]. Since the Eirreducible representations are two-dimensional, they have two basis functions each. It is clearly seen that only the E_{1u} irreducible representation exhibits a net dipole moment and will thus interact with incident light. This means all plasmon modes observed in experiments would belong to the E_{1u} mode and the mode profiles can be represented by linear combinations of the basis functions shown in Figure 11(a). To probe the plasmon modes in a heptamer, Mirin et al. calculated the absorption and scattering spectra of a silver nanoparticle heptamer with various sizes. As shown in Figure 11(b) and (c), in the smallest heptamer where the individual nanoparticles have a radius of 10 nm and are separated by 1 nm, three lowest energy peaks are observed in agreement with the plasmon hybridization model which predicts two bright modes at 340 nm and 380 nm and a dark mode at 450 nm. The bright modes are prominent in both the absorption and scattering spectra while the dark mode is weak and only visible in the absorption spectrum. Numerical simulations showed that the bright mode has a field profile indicating that the dipole moment of the central nanoparticle is aligned to that of the hexamer while the dark mode has the two dipoles anti-parallel, resulting in a small net dipole moment. As the particle sizes and gaps are increased proportionally, all modes exhibit red shifts due to the retardation effect and thus deviate from the results of plasmon hybridization which is valid only in the electrostatic regime. The retardation effect also broadens the peaks, particularly the bright modes, leading to a spectral overlap between the lowest bright mode and the dark mode. This results in the Fano resonance with its distinct asymmetric lineshape clearly visible in the scattering spectra for silver nanoparticle radii of 20 ~ 40 nm. For radius of 50 nm, the energies of the bright and dark modes become almost the same. In the case, the Fano resonance presents itself as an antiresonance with a symmetric dip in the scattering spectrum.

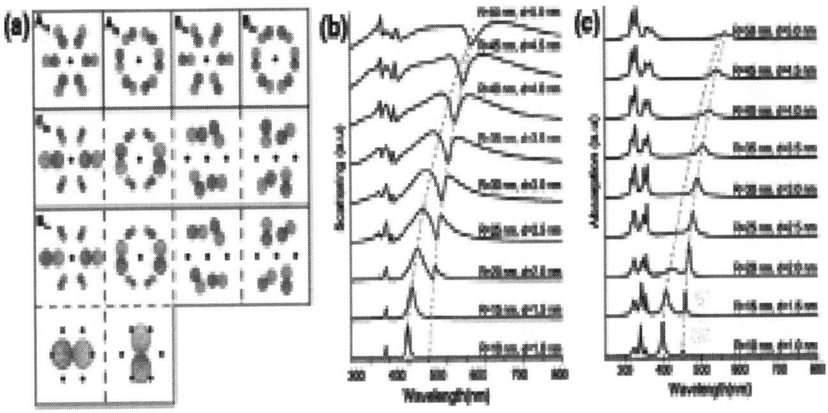

Figure 11: Silver nanoparticle heptamer.(a) Symmetry adapted basis functions of heptamer in-plane modes belonging to the *D6h* point group. (b)Scattering and (c) absorption spectra calculated with the finite element method for heptamers with various sizes. The dashed curves show the red-shift of the dark (green) and bright (blue) modes with increasing size. (From Ref. [115]) Silver nanoparticle heptamer.(a) Symmetry adapted basis functions of heptamer in-plane modes belonging to the *D6h* point group. (b) Scattering and (c) absorption spectra calculated with the finite element method for heptamers with various sizes. The dashed curves show the red-shift of the dark (green) and bright (blue) modes with increasing size (From Ref. [115]).

The effect of symmetry lowering on the Fano resonance is of interest to both the fundamental physics and potential applications. Cui et al. fabricated gold heptamers on a flexible poly(dimethylsiloxane) (PDMS) membrane and applied mechanical stress while monitoring the evolution of the Fano resonance [21]. As shown in Figure 12, the Fano dip exhibited a blue shift with increasing mechanical stress for the polarization along the direction of stress while it shifted to lower frequencies for the perpendicular polarization. Also, for the parallel polarization, an additional Fano dip shows up at large mechanical strain values. Figure 12 also shows scattering spectra obtained by FEM simulations which agree very well with the experimental spectra. The same behavior was also confirmed by the generalized multiparticle Mie calculations following the theory described in Section 2.2. To gain insight into the observed behavior, the eigenmodes of the heptamer structure were obtained by solving the boundary integral eigenvalue equations for the charge distribution, as described in Section 2.4. Note that this technique is valid only in the static limit and thus the obtained plasmon energies are not the same as the experimentally observed peak positions. Nevertheless, this study allows for the analysis of the eigenmodes with the group representation theory and thus correlate the symmetry of the system with that of the eigenmodes. As discussed earlier, the heptamer structure has the symmetry of point group D_{6h} and, among the irreducible representations of D_{6h}, E_{1u} is the only one with a net dipole moment and thus optically active. Furthermore, we only consider two lowest energy E_{1u} modes in the unstressed gold heptamer structure because all higher modes are masked by strong absorption by gold. In the first column of Figure 13, we show the charge distribution of the two lowest energy E_{1u} modes in an unstressed gold heptamer structure composed of seven identical gold spheres where the sphere diameter is 150 nm and gap between the spheres is 25 nm. Here the modes shown in Figure 13(a) and (b) belong to the lowest energy E_{1u} mode and (d) and (e) to the second lowest E_{1u} mode. The two-diomensional E_{1u} irreducible representation represetns a doubly degenerate mode with two orthogonal states having net dipole moment in the x and y directions, respectively. Accordingly, the four charge distributions shown in the first column of Figure 13 possess net dipole moment where (a) and (d) are x-dipoles and (b) and (e) are y-dipoles. Depending on the energy and the relative alignment of dipole moment of the center sphere to those of the six satelite spheres, the E_{1u} modes can be classified as dark or bright modes. The lower energy E_{1u} mode shown in Figure 13(a) and (b) is a dark mode where the dipole moment of center particle aligns against the dipole moments of satelite particles, making the total dipole moment small. On the other hand, the higher energy E_{1u} mode shown in Figure 13(d) and (e) is a bright mode where the dipole moments align together and add. The energies of these two E_{1u} modes were found to be 2.394 eV and 2.459 eV or 517.8 nm and 504.0 nm, respectively. These mode energy values would be accurate only for heptamers made of very small nanoparticles, as the boundary integral method is valid in the static limit only. For larger sizes, retardation effect will shift and broaden the modes. The resultant overlap and interference between the two modes lead to the Fano resonance. The bright mode will broaden much more significantly than the dark mode, resulting in Fano resonance which manifests itself in the form of a dip in the extinction spectrum as observed in Figure 12.

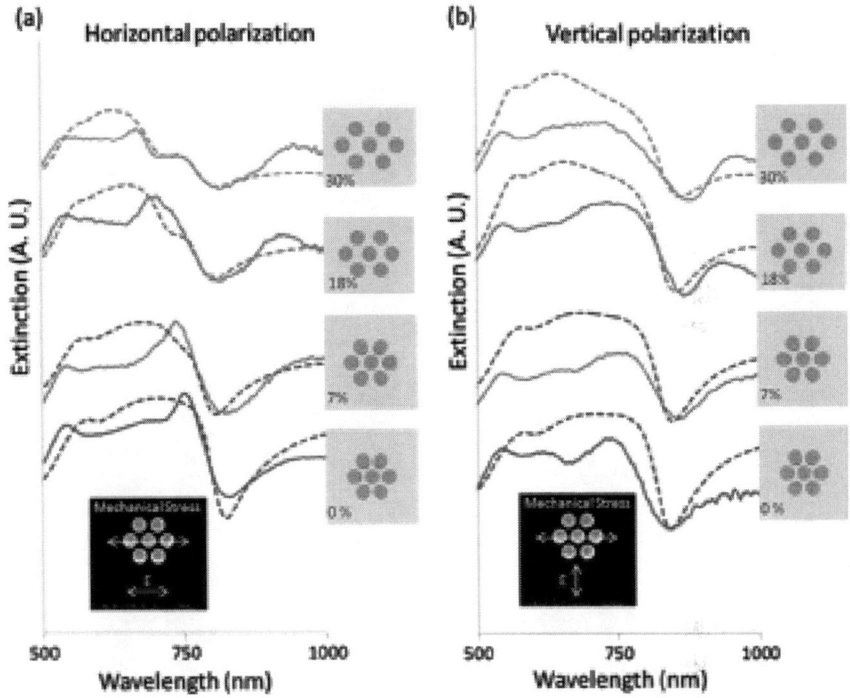

Figure 12: Mechanical tuning of gold nanoparticle heptamer.(a)-(b)show the experimentally measured and simulated extinction spectra for mechanical stress along horizontal direction. Stretching direction and polarizations are shown in the insets. Solid line shows experiment results and dashed line shows simulation results. (From Ref. [21]) Mechanical tuning of gold nanoparticle heptamer.(a)-(b) show the experimentally measured and simulated extinction spectra for mechanical stress along horizontal direction. Stretching direction and polarizations are shown in the insets. Solid line shows experiment results and dashed line shows simulation results (From Ref. [21]).

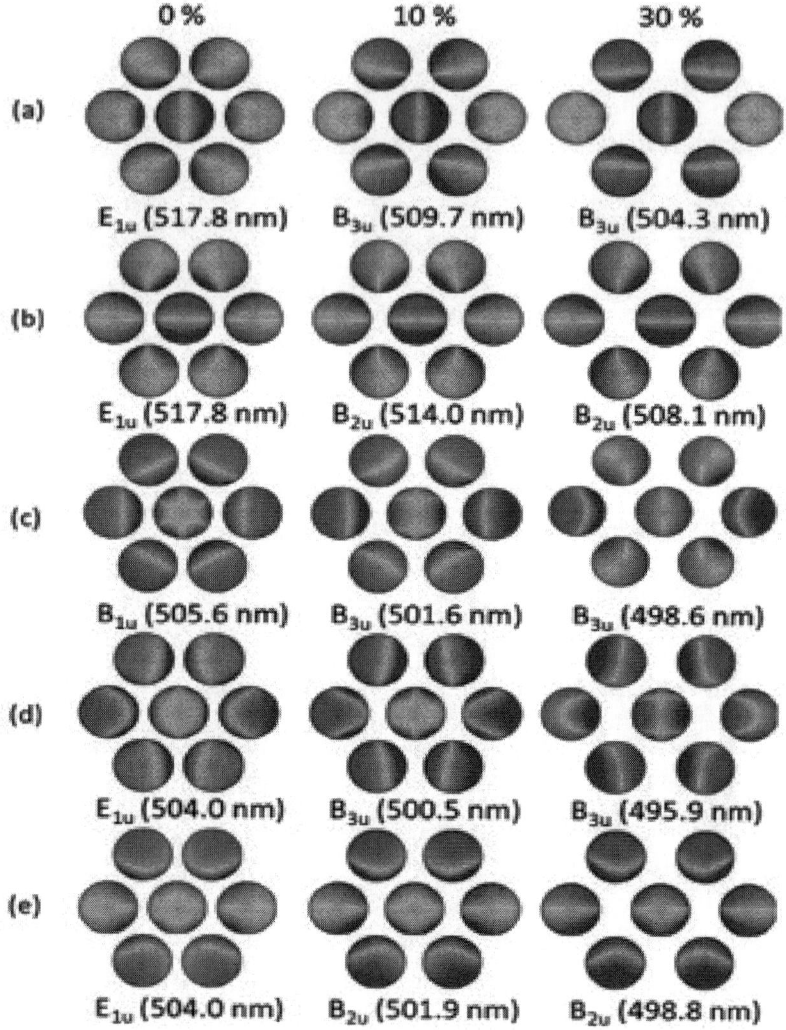

Figure 13: The evolution of charge distribution with uniaxial stress along the x direction. The first column shows the two lowest E_{1u} modes of unstressed heptamer and the B_{1u} mode. Second and third columns show B_{2u} and B_{3u} modes of the heptamer under 10% and 30% strains, respectively. The resonance wavelengths of the heptamers are also shown. (From Ref. [21]) The evolution of charge distribution with uniaxial stress along the x direction. The first column shows the two lowest E_{1u} modes of unstressed heptamer and the B_{1u} mode. Second and third columns show B_{2u} and B_{3u} modes of the heptamer under 10% and 30% strains, respectively. The resonance wavelengths of the heptamers are also shown in paraenthesis (From Ref. [21]).

When the heptamer is subject to a uniaxial mechanical stress, the symmetry of the system is lowered to D_{2h}. The doubly degenerate E_{1u} mode splits into two non-degenerate modes belonging to B_{2u} and B_{3u} irreducible representations of the point group D_{2h}. Figure 13 shows the evolution of charge distribution as the mechanical stress is applied along the x direction. It clearly shows the original doubly degenerate modes split into x-dipole (B_{3u}) and y-dipole (B_{2u}) modes. Remarkably, the nature of the modes are preserved. That is, the bright E_{1u} mode splits into bright B_{2u} and B_{3u} modes while the dark E_{1u} mode spawns dark B_{2u} and B_{3u} modes. Also, all modes shift to shorter wavelengths with increasing mechanical strain values. However, the B_{3u} modes which have dipole moment along the direction of mechanical stress shift more than the B_{2u} modes with dipole moment perpendicular to the mechanical stress. This leads to the distinct polarization dependence observed in Figure 12 as the B_{3u} modes interact with x-polarized light and B_{2u} with y-polarized light. Therefore, as the heptamer is stretched along the x direction, x-polairzed light would show resonance features at shorter wavelengths than the y-polarized light. Even when the retardation effects are included and the resonance peaks broaden and shift, this general behavior survives and leads to the experimental observation in Figure 12: the dip in the extinction spectrum due to the Fano resonance blue shifts for polarization parallel to the direction of mechanical stress but red shifts for polarization perpendicular to it. The exact numerical simulations by FEM and generalized multiparticle Mie theory also confirmed this.

In addition to the splitting of E_{1u} modes leading to the poalrization dependence, the scattering spectra for x polarization also show an additional dip at shorter wavelengths. The additional dip is apparent in the spectra for 30% and 45% strain for x polarization but is clearly missing in all spectra for y polarization. The origin of this second dip can be found by the group theoretical analysis. Briefly reiterating, in the unstressed heptamer structure possessing D_{6h} symmetry, only the E_{1u} modes have non-zero dipole moments and thus are optically active. When the symmetry is lowered to D_{2h} by uniaxial stress, the E_{1u} modes split into B_{2u} and B_{3u} modes which interact with x- and y-polarized light, respectively. Since the unstressed heptamer has two E_{1u} modes in the frequency range of interest, we obtain two B_{2u} and two B_{3u} modes, producing Fano resonance just as in the original unstressed heptamer. However, what is missing in this narrative is that the optically inactive

B_{1u} mode in the unstressed heptamer becomes optically active B_{3u} mode under uniaxial stress along the x direction. As shown in Figure 13(c), the charge distribution reveals that this B_{3u} mode is also a dark mode where the dipole moment of the center sphere aligns anti-parallel against those of the satelite spheres, thereby producing a second Fano dip in the scattering spectra. In constrast, there are no other modes of the unstressed heptamer evolving into B_{2u}mode within the frequency range we investigated and thus we do not see any additional dip for the y polarization.

Now we can take a step further and study which dipole components contribute to the formation of Fano resonance. For this, we consider nanorod heptamers composed of a circular center particle surrounded by six nanorods [22]. A circular nanoparticle supports a degenerate dipole mode with two orthogonal dipole moments having the same energy. The anisotropy of nanorod, however, lifts this degeneracy and the transverse mode has a substantially different energy than the longitudinal mode. This allows us to probe how the each dipole components hybridize separately. This design principle is schematically shown in Figure 14(a). The doubly degenerate dipole modes of the circular particle can be arranged into azimuthal dipoles and radial dipoles according to the six-fold rotational symmetry of the heptamer structure. We then replace the circular satellite particles with nanorods arranged azimuthally and radially. By designing the nanorod dimensions so that only the longitudinal modes are in the frequency range of interest, we can construct heptamers with only the azimuthal dipoles and radial dipoles. Subsequent numerical simulations revealed that the azimuthal nanorod heptamer exhibits a strong and well-defined Fano resonance (Figure 14(b)) but the radial nanorod heptamer does not support Fano resonance (Figure 14(c)). We then conducted a series of numerical simulations for a wide range of structural parameters to find that the azimuthal nanorod heptamers always support Fano resonance but the radial nanorod heptamers do not. These results indicate that the Fano resonance observed in circular heptamers are the result of hybridization of azimuthal dipole modes of the satellite particles with the center particle and that the radial dipole modes do not contribute.

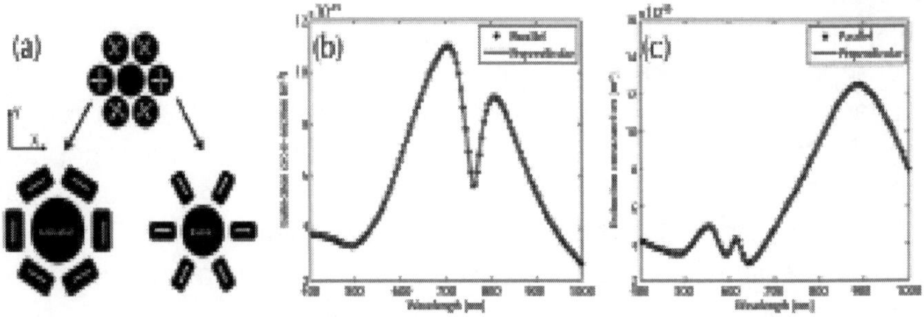

Figure 14: Gold nanorod heptamers.(a) Schematic showing the splitting of doubly degenerate modes in circular heptamer to azimuthal and radial modes in nanorod heptamer. The white colored arrows represent the radial modes and gold colored arrows represent azimuthal modes. Simulated extinction spectra for (a) azimuthal and (b)radial nanorod heptamers for the same structural parameters: individual nanorods have a length of 95 nm, width of 50 nm and thickness of 40 nm. The center particle has a diameter of 147 nm and gold thickness of 40 nm. The gaps between the center particle and the nanorods are 22 nm. (From Ref. [22]) Gold nanorod heptamers.(a) Schematic showing the splitting of doubly degenerate modes in circular heptamer to azimuthal and radial modes in nanorod heptamer. The white colored arrows represent the radial modes and gold colored arrows represent azimuthal modes. Simulated extinction spectra for (b) azimuthal and (c) radial nanorod heptamers for the same structural parameters: individual nanorods have a length of 95 nm, width of 50 nm and thickness of 40 nm. The center particle has a diameter of 147 nm and gold thickness of 40 nm. The gaps between the center particle and the nanorods are 22 nm (From Ref. [22]).

The Fano resonance supported by the azimuthal nanorod heptamer was further investigated for their behavior under mechanical stress and symmetry lowering. As before, the mechanical stress was applied along the *x* (horizontal) direction. Figure 15 shows the simulated extinction and absorption spectra at the induced mechanical strain values of 0%, 13%, 26%, and 42%. For the polarization parallel to the mechanical stress (*x* direction), the Fano resonance exhibited a moderate blue shift while for the perpendicular polarization it showed a small blue shift. In addition to the spectral shifts with applied mechanical stress, the absorption spectra for the perpendicular polarization showed an additional feature at shorter wavelengths when the heptamer was under mechanical stress. This feature located near 620 nm was particularly noticeable at high strain values of 26% and 42%. The behavior of the spectral features in Figure 15 can be explained by analyzing the eigenmodes obtained by the boundary integral method. The evolution of charge distribution is dictated mainly by the symmetry reduction from D_{6h} to D_{2h} and is similar to that observed in the circular heptamer under mechanical stress described earlier. As was the case in the circular heptamer, the doubly degenerate E_{1u} mode of the azimuthal nanorod heptamer splits into two non-degenerate modes belonging to B_{2u} (*y*-dipole) and B_{3u} (*x*-dipole) irreducible representations of the point group D_{2h}. Furthermore, the bright E_{1u} mode was found to split into bright B_{2u} and B_{3u} modes and the dark E_{1u} mode into dark B_{2u} and B_{3u} modes. Both the dark and bright B_{2u} and B_{3u} modes shift to shorter wavelengths with increasing mechanical strain values. However, the B_{3u} modes, which have dipole moment along the direction of mechanical stress, shifts more than the B_{2u} modes whose dipole moment is perpendicular to the mechanical stress, leading to the polarization dependence observed in the extinction and absorption spectra in Figure 15. Once again, the retardation effect results in overlap and interference between the dark and bright modes thus causing the Fano resonance to be observed as a dip in the extinction spectrum.

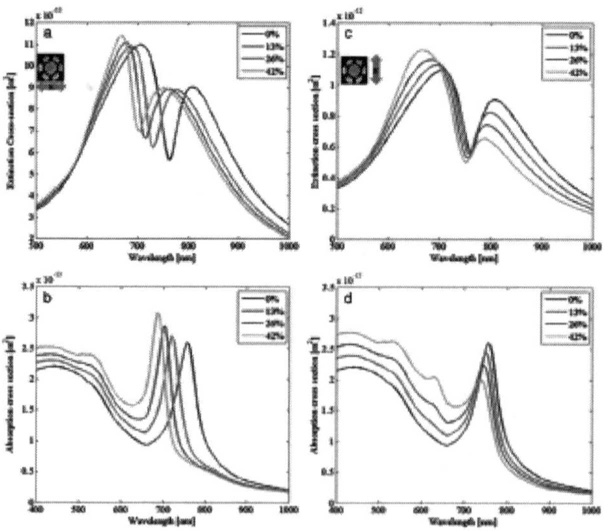

Figure 15: Simulated extinction and absorption spectra for the azimuthal nanorod heptamer for different induced mechanical strain values of 0%,

13%, 26% and 42%. (a) and (c) plot the extinction spectra for parallel and perpendicular polarizations (orientations in inset) respectively while (b) and (d) plot the absorption spectra for parallel and perpendicular polarizations respectively. (From Ref. [22]) Simulated extinction and absorption spectra for the azimuthal nanorod heptamer for different induced mechanical strain values of 0%, 13%, 26% and 42%.(a) and (c) plot the extinction spectra for parallel and perpendicular polarizations (orientations in inset) respectively while (b) and (d) plot the absorption spectra for parallel and perpendicular polarizations respectively (From Ref. [22]).

The additional absorption peak observed under stress can be explained by noting that the optically inactive B_{2u} mode in the unstressed heptamer becomes an optically active B_{2u} mode under uniaxial stress along the *x* direction. Here we have an unfortunate coincidence of having the same label for two very different modes. The B_{2u} irreducible representation in D_{6h} symmetry has no net dipole moment and thus represents an optically inactive mode. However, the B_{2u} irreducible representation in D_{2h} symmetry has a finite dipole moment along the *y*-direction and thus represents an optically active mode that can interact with *y*-polarized light. The energy of this B_{2u} mode is higher than the lowest dark B_{2u} mode originating from the dark E_{1u} mode of unstressed heptamer, thereby producing an additional feature at a higher energy. The azimuthal nanorod heptamer differs from the circular heptamer in the behavior of this optically active mode. In the circular heptamer, the optically inactive B_{1u} mode was located between the two lowest energy E_{1u} modes in the unstressed heptamer and became optically active B_{3u} mode under uniaxial stress along the *x* direction. As a result, in the circular heptamer, the additional mode was observed for x-polarized light. The difference between the circular heptamer and azimuthal nanorod heptamer is readily understood by comparing the eigenmode charge distributions. In the B_{1u} mode observed in the unstressed circular heptamer, the satellite particles have their dipole moments aligned along the radial direction (Figure 13c). In the azimuthal nanorod heptamer, this configuration would have had a much higher energy because the radial dimensions of the nanorods are much smaller than the circular particles used in circular heptamer. On the other hand, the B_{2u} mode in which the satellite particles have their dipole moments aligned along the azimuthal direction has a lower energy because the nanorods have larger dimensions along that direction. For these reasons, we find the B_{2u} mode is located between the two lowest energy E_{1u} modes in the azimuthal nanorod heptamer structure. These behaviors predicted by the theoretical modeling studies have been confirmed by experiments on gold nanorod heptamers fabricated by electron-beam lithography [22].

CONCLUSIONS AND OUTLOOK

Plasmonics in mutually interacting nanostructures is an exciting field with rich physics and numerous potential applications. The recent progress clearly demonstrates that plasmonic nanostructures allow us to precisely control the nanoscale optical modes and how they interact with one another. These capabilities make the coupled plasmonic nanostructures an ideal platform to study nanoscale optical phenomena and build optical devices with novel functionalities. Though fundamentally a quantum mechanical entity, surface plasmons can be described well by classical electrodynamics in most cases. The theoretical framework presented in this review provides a firm foundation for designing interesting structures and analyzing their properties. Also, the strong field enhancement near the metal surface and the complex field profiles resulting from the interaction among the nanostructures makes the optical properties highly sensitive to the environment and consequently leads to exotic properties such as perfect absorption, invisibility and Fano resonance.

Some of the latest developments in the coupled plasmonic systems deal specifically with the quantum mechanical nature of the phenomenon. In surface enhanced Raman scattering and many other applications, it is generally advantageous to achieve high local field strength. In the simplest geometry of a dimer, a smaller gap supports higher field. However, when the two nanoparticles in a dimer are nearly touching, the electrons may tunnel through the gap, resulting in a dramatic decrease of local field strength and shift of plasmon energy [26,116]. Furthermore, at such a small length scale, the abrupt interface between two materials is unrealistic and the non-locality of the dielectric function has to be considered [117]. These considerations have led to the development of a quantum-corrected model that could correctly reproduce the fully quantum mechanical calculations [118]. Unveiling the quantum mechanical nature of the gap plasmons in extremely small gaps remains an active research field with hopes of discovering new physics and also with potential for developing novel quantum devices.

Another interesting subject of fundamentally quantum nature is the coupling between plasmon and atomic or molecular exciton. The interaction between plasmon and exciton is of great fundamental interest with strong technological implications in organic electronics and photovoltaics. It has been shown that the surface plasmon can be used to control the relaxation pathways of molecular excitons [119]. Furthermore, in a fashion similar to plasmon hybridization, atomic or molecular exciton can hybridize with plasmon, forming a new quantum mechanically entity sometimes called plexciton [120,121]. The strong coupling between plasmon and exciton has recently been observed in an organic photovoltaic material [122]. The strong interaction could result in dramatic changes in excitation and relaxation dynamics of the molecular excitons, opening new opportunities in controlling and engineering the optical processes in organic materials.

In closing, surface plasmons provide a powerful means to gain access to the near field components that could result in a wide range of unconventional properties. Controlling and manipulating the mutual interactions among the plasmonic nanostructures offer design flexibility and engineering freedom unparalleled by any other photonics technologies. Obtaining high quality materials with low loss in precise nanoscale geometry will continue to be a challenge but the remarkable progress in nanofabrication technologies in the past decades makes it a perfect time to invest in research on the plasmonic nanostructures.

REFERENCES

1. DK Gramotnev, SI Bozhevolnyi, plasmonics beyond the diffraction limit. Nat. Photon. 4(2), 83–91 (2010)
2. J-S Lee, MS Han, CA Mirkin, Colorimetric Detection of Mercuric Ion (Hg2+) in Aqueous Media using DNA-Functionalized Gold Nanoparticles. Angew. Chem. 119(22), 4171–4174 (2007)
3. JJ Storhoff, R Elghanian, RC Mucic, CA Mirkin, RL Letsinger, One-pot colorimetric differentiation of polynucleotides with single base imperfections using gold nanoparticle probes. J. Am. Chem. Soc. 120(9), 1959–1964 (1998)

4. J Chen, F Saeki, BJ Wiley, H Cang, MJ Cobb, Z-Y Li, L Au, H Zhang, MB Kimmey, Li, Y Xia, Gold Nanocages: Bioconjugation and Their Potential Use as Optical Imaging Contrast Agents. Nano Lett. 5(3), 473–477 (2005)
5. EB Dickerson, EC Dreaden, X Huang, IH El-Sayed, H Chu, S Pushpanketh, JF McDonald, MA El-Sayed, Gold nanorod assisted near-infrared plasmonic photothermal therapy (PPTT) of squamous cell carcinoma in mice. Cancer Lett. 269(1), 57–66 (2008)
6. KY Lin, AF Bagley, AY Zhang, DL Karl, SS Yoon, SN Bhatia, Gold nanorod photothermal therapy in a genetically engineered mouse model of soft tissue sarcoma. Nano LIFE 1(3), 277–287 (2010)
7. S Nie, SR Emory, Probing single molecules and single nanoparticles by surfaceenhanced Raman scattering. Science 275(5303), 1102–1106 (1997)
8. AM Michaels, M Nirmal, LE Brus, Surface Enhanced Raman Spectroscopy of Individual Rhodamine 6G Molecules on Large Ag Nanocrystals. J. Am. Chem. Soc. 121(43), 9932–9939 (1999)
9. JR Lakowicz, Radiative decay engineering 5: metal-enhanced fluorescence and plasmon emission. Anal. Biochem. 337(2), 171–194 (2005)
10. P Andrew, Energy Transfer Across a Metal Film Mediated by Surface Plasmon Polaritons. Science 306(5698), 1002–1005 (2004)
11. M Lunz, VA Gerard, YK Gun'ko, V Lesnyak, N Gaponik, AS Susha, AL Rogach, AL Bradley, Surface Plasmon Enhanced Energy Transfer between Donor and Acceptor CdTe Nanocrystal Quantum Dot Monolayers. Nano Lett. 11(8), 3341–3345 (2011)
12. L Novotny, N van Hulst, Antennas for light. Nat. Photon. 5(2), 83–90 (2011) 54
13. E Prodan, C Radloff, NJ Halas, P Nordlander, A hybridization model for the plasmon response of complex nanostructures. Science 302(5644), 419–422 (2003)
14. P Nordlander, C Oubre, E Prodan, K Li, MI Stockman, Plasmon hybridization in nanoparticle dimers. Nano Lett. 4(5), 899–903 (2004)
15. VK Lin, SL Teo, R Marty, A Arbouet, C Girard, E Alarcon-Llado, SH Liu, MY Han, S Tripathy, A Mlayah, Dual wavelength sensing

based on interacting gold nanodisk trimers. Nanotechnology 21(30), 305501 (2010)

16. AJ Mastroianni, SA Claridge, AP Alivisatos, Pyramidal and chiral groupings of gold nanocrystals assembled using DNA scaffolds. J. Am. Chem. Soc. 131(24), 8455–8459 (2009)
17. DE Gómez, KC Vernon, TJ Davis, Symmetry effects on the optical coupling between plasmonic nanoparticles with applications to hierarchical structures. Phys. Rev. B 81(7), 075414 (2010)
18. M Hentschel, M Saliba, R Vogelgesang, H Giessen, AP Alivisatos, N Liu, Transition from Isolated to Collective Modes in Plasmonic Oligomers. Nano Lett. 10(7), 2721–2726 (2010)
19. JB Lassiter, H Sobhani, JA Fan, J Kundu, F Capasso, P Nordlander, NJ Halas, Fano Resonances in Plasmonic Nanoclusters: Geometrical and Chemical Tunability. Nano Lett.10(8), 3184–3189 (2010)
20. K Bao, NA Mirin, P Nordlander, Fano resonances in planar silver nanosphere clusters. Appl. Phys. A 100(2), 333–339 (2010)
21. Y Cui, J Zhou, VA Tamma, W Park, Dynamic Tuning and Symmetry Lowering of Fano Resonance in Plasmonic Nanostructure. ACS Nano 6(3), 2385–2393 (2012)
22. VA Tamma, Y Cui, J Zhou, W Park, Nanorod orientation dependence of tunable Fano resonance in plasmonic nanorod heptamers. Nanoscale 5(4), 1592–1602 (2013)
23. M Tinkham, *Group Theory and Quantum Mechanics* (Dover Publications, Mineola, New York, 2003)
24. F Flores, F Garcia-Moliner, Model-independent theory of surface plasmons. Solid State Commun. 11(9), 1295–1298 (1972)
25. G Barton, Some surface effects in the hydrodynamic model of metals. Rep. Prog. Phys.42(6), 963 (1979)
26. J Zuloaga, E Prodan, P Nordlander, Quantum Description of the Plasmon Resonances of a Nanoparticle Dimer. Nano Lett. 9(2), 887–891 (2009)
27. Z Jacob, VM Shalaev, Plasmonics Goes Quantum. Science 334(6055), 463–464 (2011)
28. DJ Bergman, MI Stockman, Surface plasmon amplification by stimulated emission of radiation: quantum generation of coherent

surface plasmons in nanosystems. Phys. Rev. Lett. 90(2), 027402 (2003)

29. A Archambault, F Marquier, J-J Greffet, C Arnold, Quantum theory of spontaneous and stimulated emission of surface plasmons. Phys. Rev. B 82(3), 035411 (2010)
30. RH Ritchie, HB Eldridge, Optical emission from irradiated foils. I. Phys. Rev. 126(6), 1935 (1962)
31. J Dionne, L Sweatlock, H Atwater, A Polman, Planar metal plasmon waveguides: frequency-dependent dispersion, propagation, localization, and loss beyond the free electron model. Phys. Rev. B 72(7), 075405 (2005)
32. J Dionne, L Sweatlock, H Atwater, A Polman, Plasmon slot waveguides: Towards chip-scale propagation with subwavelength-scale localization. Phys. Rev. B 73(3), 035407 (2006) 55
33. R Zia, M Selker, M Brongersma, Leaky and bound modes of surface plasmon waveguides. Phys. Rev. B 71(16), 165431 (2005)
34. T Holmgaard, S Bozhevolnyi, Theoretical analysis of dielectric-loaded surface plasmonpolariton waveguides. Phys. Rev. B 75(24), 245405 (2007)
35. G Mie, Beiträge zur Optik trüber Medien, speziell kolloidaler Metallösungen. Ann. Phys.330(3), 377–445 (1908)
36. YL Xu, Electromagnetic scattering by an aggregate of spheres. Appl. Opt. 34(21), 4573–4588 (1995)
37. Y-L Xu, Electromagnetic scattering by an aggregate of spheres: far field. Appl. Opt. 36(36), 9496–9508 (1997)
38. Y-L Xu, G BÅ, Comparison between multisphere light-scattering calculations: rigorous solution and discrete-dipole approximation. Astrophys. J. 513(2), 894 (1999)
39. CF Bohren, DR Huffman, *Absorption and Scattering of Light by Small Particles* (JWILEY-VCH Verlag GmbH & Co. KGaA, Weinheim, 2008)
40. N Stefanou, V Karathanos, A Modinos, Scattering of electromagnetic waves by periodic structures. J. Phys. Condens. Matter 4(36), 7389 (1992)
41. X Wang, XG Zhang, Q Yu, B Harmon, Multiple-scattering theory for electromagnetic waves. Phys. Rev. B 47(8), 4161–4167 (1993)

42. A Moroz, Density-of-states calculations and multiple-scattering theory for photons. Phys. Rev. B 51(4), 2068 (1995)
43. N Stefanou, A Modinos, Scattering of light from a two-dimensional array of spherical particles on a substrate. J. Phys. Condens. Matter 3(41), 8135 (1991)
44. N Stefanou, V Yannopapas, A Modinos, Heterostructures of photonic crystals: frequency bands and transmission coefficients. Comput. Phys. Commun. 113(1), 49–77 (1998)
45. JB Pendry, Photonic Band Structures. J. Mod. Opt. 41(2), 209–229 (1994)
46. JCM Garnett, Colours in Metal Glasses and in Metallic Films. Philos. Trans. R. Soc. A Math. Phys. Eng. Sci. 203(359), 385–420 (1904)
47. WT Doyle, Optical properties of a suspension of metal spheres. Phys. Rev. B 39(14), 9852 (1989)
48. V Yannopapas, Negative refraction in random photonic alloys of polaritonic and plasmonic microspheres. Phys. Rev. B 75(3), 035112 (2007)
49. VA Tamma, JH Lee, Q Wu, W Park, Visible frequency magnetic activity in silver nanocluster metamaterial. Appl. Opt. 49(7), A11–A17 (2010)
50. CF Bohren, Applicability of effective-medium theories to problems of scattering and absorption by nonhomogeneous atmospheric particles. J. Atmos. Sci. 43, 468–475 (1986)
51. S Yoo, Q-H Park, Effective permittivity for resonant plasmonic nanoparticle systems via dressed polarizability. Opt. Express 20(15), 16480–16489 (2012)
52. TC Choy, *Effective Medium Theory* (Oxford University Press, New York, 1999)
53. P Berini, Plasmon-polariton waves guided by thin lossy metal films of finite width: Bound modes of symmetric structures. Phys. Rev. B 61(15), 10484 (2000)
54. P Berini, Plasmon-polariton waves guided by thin lossy metal films of finite width: Bound modes of asymmetric structures. Phys. Rev. B 63(12), 125417 (2001)

55. H Wang, DW Brandl, P Nordlander, NJ Halas, Plasmonic Nanostructures: Artificial Molecules. Acc. Chem. Res. 40(1), 53–62 (2007)
56. JA Fan, K Bao, C Wu, J Bao, R Bardhan, NJ Halas, VN Manoharan, G Shvets, P Nordlander, F Capasso, Fano-like Interference in Self-Assembled Plasmonic Quadrumer 56 Clusters. Nano Lett. 10(11), 4680–4685 (2010)
57. I Mayergoyz, D Fredkin, Z Zhang, Electrostatic (plasmon) resonances in nanoparticles. Phys. Rev. B 72(15), 155412 (2005)
58. VM Shalaev, Optical negative-index metamaterials. Nat. Photon. 1(1), 41–48 (2007)
59. W Park, J Kim, Negative-Index Materials: Optics by Design. MRS Bull. 33(10), 907–914 (2008)
60. Z-P Yang, L Ci, JA Bur, S-Y Lin, PM Ajayan, Experimental Observation of an Extremely Dark Material Made By a Low-Density Nanotube Array. Nano Lett. 8(2), 446–451 (2008)
61. N Landy, S Sajuyigbe, J Mock, D Smith, W Padilla, Perfect Metamaterial Absorber. Phys. Rev. Lett. 100(20), 207402 (2008)
62. D Lu, E Rengnath, Y Cui, Z Wang, Y Ding, W Park, Interaction of two plasmon modes in the organic photovoltaic devices with patterned back-electrode. Appl. Phys. Lett. 102(24), 241114 (2013)
63. V Kravets, F Schedin, A Grigorenko, Plasmonic blackbody: Almost complete absorption of light in nanostructured metallic coatings. Phys. Rev. B 78(20), 205405 (2008)
64. JD Jackson, *Classical Electrodynamics* (Wiley, New York, 1998)
65. FJ García-Vidal, JM Pitarke, JB Pendry, Effective medium theory of the optical properties of aligned carbon nanotubes. Phys. Rev. Lett. 78(22), 4289 (1997)
66. VG Kravets, S Neubeck, AN Grigorenko, AF Kravets, Plasmonic blackbody: Strong absorption of light by metal nanoparticles embedded in a dielectric matrix. Phys. Rev. B81(16), 165401 (2010)
67. EFC Driessen, MJA de Dood, The perfect absorber. Appl. Phys. Lett. 94(17), 171109 (2009)

68. C Hägglund, SP Apell, B Kasemo, Maximized Optical Absorption in Ultrathin Films and Its Application to Plasmon-Based Two-Dimensional Photovoltaics. Nano Lett. 10(8), 3135–3141 (2010)
69. C Hägglund, SP Apell, Resource efficient plasmon-based 2D-photovoltaics with reflective support. Opt. Express 18(103), 343–356 (2010)
70. C Hägglund, G Zeltzer, R Ruiz, I Thomann, H-B-R Lee, ML Brongersma, SF Bent, Self-Assembly Based Plasmonic Arrays Tuned by Atomic Layer Deposition for Extreme Visible Light Absorption. Nano Lett. 13(7), 3352–3357 (2013)
71. M Svedendahl, P Johansson, M Käll, Complete Light Annihilation in an Ultrathin Layer of Gold Nanoparticles. Nano Lett. 13(7), 3053–3058 (2013)
72. A Alù, N Engheta, Achieving transparency with plasmonic and metamaterial coatings. Phys. Rev. E. 72(1), 016623 (2005)
73. A Alù, N Engheta, Multifrequency Optical Invisibility Cloak with Layered Plasmonic Shells. Phys. Rev. Lett. 100(11), 113901 (2008)
74. A Alù, N Engheta, Cloaking a Sensor. Phys. Rev. Lett. 102(23), 233901 (2009)
75. A Alù, N Engheta, Cloaked Near-Field Scanning Optical Microscope Tip for Noninvasive Near-Field Imaging. Phys. Rev. Lett. 105(26), 263906 (2010)
76. F Bilotti, S Tricarico, F Pierini, L Vegni, Cloaking apertureless near-field scanning optical microscopy tips. Opt. Lett. 36(2), 211–213 (2011)
77. S Tricarico, F Bilotti, L Vegni, Reduction of optical forces exerted on nanoparticles covered by scattering cancellation based plasmonic cloaks. Phys. Rev. B 82(4), 045109 (2010)
78. VA Tamma, Y Cui, W Park, Scattering reduction at near-infrared frequencies using plasmonic nanostructures. Opt. Express 21(1), 1041–1056 (2013) 57
79. DE Aspnes, Local-field effects and effective-medium theory: A microscopic perspective. Am. J. Phys. 50, 704 (1982)
80. S Mühlig, A Cunningham, J Dintinger, M Farhat, SB Hasan, T Scharf, T Bürgi, F Lederer, C Rockstuhl, A self-assembled three-dimensional cloak in the visible. Sci. Rep. 3, 2328 (2013)

81. KL Young, MB Ross, MG Blaber, M Rycenga, Using DNA to Design Plasmonic Metamaterials with Tunable Optical Properties. Adv. Mater. 26(4), 653–659 (2014)
82. SY Park, AKR Lytton-Jean, B Lee, S Weigand, GC Schatz, CA Mirkin, DNAprogrammable nanoparticle crystallization. Nature 451(7178), 553–556 (2008)
83. D Nykypanchuk, MM Maye, D van der Lelie, O Gang, DNA-guided crystallization of colloidal nanoparticles. Nature 451(7178), 549–552 (2008)
84. CE Talley, JB Jackson, C Oubre, NK Grady, CW Hollars, SM Lane, TR Huser, P Nordlander, NJ Halas, Surface-Enhanced Raman Scattering from Individual Au Nanoparticles and Nanoparticle Dimer Substrates. Nano Lett. 5(8), 1569–1574 (2005)
85. P Muhlschlegel, HJ Eisler, OJF Martin, B Hecht, DW Pohl, Resonant optical antennas. Science308(5728), 1607–1609 (2005)
86. T Atay, J-H Song, AV Nurmikko, Strongly Interacting Plasmon Nanoparticle Pairs: From Dipole−Dipole Interaction to Conductively Coupled Regime. Nano Lett. 4(9), 1627–1631 (2004)
87. KH Su, QH Wei, X Zhang, JJ Mock, DR Smith, S Schultz, Interparticle Coupling Effects on Plasmon Resonances of Nanogold Particles. Nano Lett. 3(8), 1087–1090 (2003)
88. W Rechberger, A Hohenau, A Leitner, JR Krenn, B Lamprecht, FR Aussenegg, Optical properties of two interacting gold nanoparticles. Opt. Commun. 220(1), 137–141 (2003)
89. L Gunnarsson, T Rindzevicius, J Prikulis, B Kasemo, M Käll, S Zou, GC Schatz, Confined Plasmons in Nanofabricated Single Silver Particle Pairs: Experimental Observations of Strong Interparticle Interactions. J. Phys. Chem. B 109(3), 1079–1087 (2005)
90. P Biagioni, J-S Huang, B Hecht, Nanoantennas for visible and infrared radiation. Rep. Prog. Phys. 75(2), 024402 (2012)
91. A Taflove, SG Johnson, A Oskooi, *Advances in FDTD Computational Electrodynamics* (Artech House, Norwood, MA, 2013)
92. AC Polycarpou, Introduction to the Finite Element Method in Electromagnetics (Morgan & Claypool Publishers, 2006).

93. U Kreibig, M Vollmer, *Optical Properties of Metal Clusters* (Springer Verlag, Berlin, 2010)
94. M Gluodenis, CA Foss, The Effect of Mutual Orientation on the Spectra of Metal Nanoparticle Rod–Rod and Rod–Sphere Pairs. J. Phys. Chem. B 106(37), 9484–9489 (2002)
95. PK Jain, W Huang, MA El-Sayed, On the Universal Scaling Behavior of the Distance Decay of Plasmon Coupling in Metal Nanoparticle Pairs: A Plasmon Ruler Equation. Nano Lett. 7(7), 2080–2088 (2007)
96. PK Jain, MA El-Sayed, Universal Scaling of Plasmon Coupling in Metal Nanostructures: Extension from Particle Pairs to Nanoshells. Nano Lett. 7(9), 2854–2858 (2007)
97. PK Jain, MA El-Sayed, Surface Plasmon Coupling and Its Universal Size Scaling in Metal Nanostructures of Complex Geometry: Elongated Particle Pairs and Nanosphere 58 Trimers. J. Phys. Chem. C 112(13), 4954–4960 (2008)
98. VA Podolskiy, AK Sarychev, VM Shalaev, Plasmon modes in metal nanowires and left-handed materials. J. Nonlinear Optic. Phys. Mat. 11(1), 65–74 (2002)
99. VA Podolskiy, AK Sarychev, VM Shalaev, Plasmon modes and negative refraction in metal nanowire composites. Opt. Express 11(7), 735–745 (2003)
100. VM Shalaev, W Cai, UK Chettiar, HK Yuan, AK Sarychev, VP Drachev, AV Kildishev, Negative index of refraction in optical metamaterials. Opt. Lett. 30(24), 3356–3358 (2005)
101. S Zhang, W Fan, N Panoiu, K Malloy, R Osgood, S Brueck, Experimental Demonstration of Near-Infrared Negative-Index Metamaterials. Phys. Rev. Lett. 95(13), 137404 (2005)
102. G Dolling, C Enkrich, M Wegener, CM Soukoulis, S Linden, Simultaneous negative phase and group velocity of light in a metamaterial. Science 312(5775), 892–894 (2006)
103. AM Funston, C Novo, TJ Davis, P Mulvaney, Plasmon Coupling of Gold Nanorods at Short Distances and in Different Geometries. Nano Lett. 9(4), 1651–1658 (2009)
104. DW Brandl, NA Mirin, P Nordlander, Plasmon Modes of Nanosphere Trimers and Quadrumers. J. Phys. Chem. B 110(25), 12302–12310 (2006)

105. Z Zhang, A Weber-Bargioni, SW Wu, S Dhuey, S Cabrini, PJ Schuck, Manipulating nanoscale light fields with the asymmetric bowtie nano-colorsorter. Nano Lett. 9(12), 4505–4509 (2009)
106. M Rahmani, T Tahmasebi, Y Lin, B Lukiyanchuk, TYF Liew, MH Hong, Influence of plasmon destructive interferences on optical properties of gold planar quadrumers. Nanotechnology22(24), 245204 (2011)
107. SJ Barrow, X Wei, JS Baldauf, AM Funston, P Mulvaney, The surface plasmon modes of self-assembled gold nanocrystals. Nat. Commun. 3, 1275 (2012)
108. (U), [http://arxiv.org/abs/cond-mat/0502210] *webcite*. Fano, "On the absorption spectrum of noble gases at the arc spectrum limit," arXiv.org, vol. cond-mat.other. 08-Feb-2005 .
109. A Hessel, AA Oliner, A new theory of Wood's anomalies on optical gratings. Appl. Opt.4(10), 1275–1297 (1965)
110. B luk'yanchuk, NI Zheludev, SA Maier, NJ Halas, P Nordlander, H Giessen, CT Chong, The Fano resonance in plasmonic nanostructures and metamaterials. Nat. Mater. 9(9), 707–715 (2010)
111. A Miroshnichenko, S Flach, Y Kivshar, Fano resonances in nanoscale structures. Rev. Mod. Phys. 82(3), 2257–2298 (2010)
112. Y Francescato, V Giannini, SA Maier, Plasmonic Systems Unveiled by Fano Resonances. ACS Nano 6(2), 1830–1838 (2012)
113. N Liu, T Weiss, M Mesch, L Langguth, U Eigenthaler, M Hirscher, C Sönnichsen, H Giessen, Planar Metamaterial Analogue of Electromagnetically Induced Transparency for Plasmonic Sensing. Nano Lett. 10(4), 1103–1107 (2010)
114. C Wu, AB Khanikaev, R Adato, N Arju, AA Yanik, H Altug, G Shvets, Fanoresonant asymmetric metamaterials for ultrasensitive spectroscopy and identification of molecular monolayers. Nat. Mater. 11(1), 69–75 (2012)
115. NA Mirin, K Bao, P Nordlander, Fano Resonances in Plasmonic Nanoparticle Aggregates †. J. Phys. Chem. A 113(16), 4028–4034 (2009)
116. J Zuloaga, E Prodan, P Nordlander, Quantum Plasmonics: Optical Properties and Tunability of Metallic Nanorods. ACS Nano 4(9), 5269–5276 (2010)

117. FJ García de Abajo, Nonlocal Effects in the Plasmons of Strongly Interacting Nanoparticles, Dimers, and Waveguides. J. Phys. Chem. C 112(46), 17983–59–17987 (2008)

118. R Esteban, AG Borisov, P Nordlander, J Aizpurua, Bridging quantum and classical plasmonics with a quantum-corrected model. Nat. Commun. 3, 825 (2012)

119. G Wiederrecht, J Hall, A Bouhelier, Control of Molecular Energy Redistribution Pathways via Surface Plasmon Gating. Phys. Rev. Lett. 98(8), 083001 (2007)

120. NT Fofang, NK Grady, Z Fan, AO Govorov, NJ Halas, Plexciton Dynamics: Exciton−Plasmon Coupling in a J-Aggregate−Au Nanoshell Complex Provides a Mechanism for Nonlinearity. Nano Lett. 11(4), 1556–1560 (2011)

121. A Manjavacas, FG de Abajo, P Nordlander, Quantum plexcitonics: strongly interacting plasmons and excitons. Nano Lett. 11(6), 2318–2323 (2011)

122. N Azarova, AJ Ferguson J, van de Lagemaat, E Rengnath, W Park, JC Johnson, Coupling between a Molecular Charge-Transfer Exciton and Surface Plasmons in a Nanostructured Metal Grating, J. Phys. Chem. Lett. 2658–2663 (2013).

Chapter 7

Tuning of Structural, Optical, and Magnetic Properties of Ultrathin and Thin ZnO Nanowire Arrays for Nano Device Applications

Satinder K Shrama[1, 2], Neelam Saurakhiya[2], Sumit Barthwal[2], Rudra Kumar[2], and Ashutosh Sharma[2]

[1]School of Computing and Electrical Engineering, Indian Institute of Technology (IIT)-Mandi, Mandi, Himanchal Pradesh 175001, India

[2]DST Unit on Nanosciences, Department of Chemical Engineering, Indian Institute of Technology (IIT)-Kanpur, Kanpur, Uttar Pradesh 208016, India

ABSTRACT

One-dimensional (1-D) ultrathin (15 nm) and thin (100 nm) aligned 1-D (0001) and (0001¯) oriented zinc oxide (ZnO) nanowire (NW) arrays were fabricated on copper substrates by one-step electrochemical deposition inside the pores of polycarbonate membranes. The aspect ratio dependence of the compressive stress because of the lattice mismatch between NW array/substrate interface and crystallite size variations is investigated. X-ray diffraction results show that the polycrystalline ZnO NWs have a wurtzite structure with $a = 3.24$ Å, $c = 5.20$ Å, and [002] elongation. HRTEM and SAED pattern confirmed the polycrystalline nature of ultrathin ZnO NWs and lattice spacing of 0.58 nm. The crystallite size and compressive stress in as-grown 15- and 100-nm wires are 12.8 nm and 0.2248 GPa and 22.8 nm and 0.1359 GPa, which changed to 16.1 nm and 1.0307 GPa and 47.5 nm and 1.1677 GPa after annealing at 873 K in ultrahigh vacuum (UHV), respectively. Micro-Raman spectroscopy showed that the increase in E_2 (high) phonon frequency corresponds to much higher compressive stresses in ultrathin NW arrays. The minimum-maximum magnetization magnitude for the as-grown ultrathin and thin NW arrays are approximately 8.45×10^{-3} to 8.10×10^{-3} emu/g and approximately 2.22×10^{-7} to 2.190×10^{-7} emu/g, respectively. The magnetization in 15-nm NW arrays is about 4 orders of magnitude higher than that in the 100 nm arrays but can be reduced greatly by the UHV annealing. The origin of ultrathin and thin NW array ferromagnetism may be the exchange interactions between localized electron spin moments resulting from oxygen vacancies at the surfaces of ZnO NWs. The n-type conductivity of 15-nm NW array is higher by about a factor of 2 compared to that of the 100-nm ZnO NWs, and both can be greatly enhanced by UHV annealing. The ability to tune the stresses and the structural and relative occupancies of ZnO NWs in a wide range by annealing has important implications for the design of advanced photonic, electronic, and magneto-optic nano devices.

BACKGROUND

One-dimensional (1-D) inorganic nanostructures have stimulated great interest because of their unique physical and chemical properties [1-4] such as flexibility of nanostructures [5-7], metal-insulator transition [4,8], superior mechanic toughness [6], higher luminescence efficiency, and lower lasing threshold [8,9]. Moreover, 1-D nanostructure research has elucidated many biomarkers [10] which have the potential to greatly improve disease diagnosis. Among these materials, zinc oxide (ZnO), which is an n-type II-VI semiconductor with wide band gap energy (E_g = 3.37 eV at 300 K) and large exciton binding energy of (60 meV), has been proven as a promising candidate for multifunctional materials [11-15], variators [16], bulk acoustic wave devices [17], magneto-optic devices, UV light-emitting devices [18], gas sensors [10,19], solar cells [11,20], and field emission display devices [15,21]. In addition to this, ZnO exhibits piezoelectricity [22] in surface acoustic wave (SAW) devices and bio-compatibility [23]. Thus, ZnO-based 1-D nanostructures are very attractive materials to explore further because of their structural, electronic, optical, and magnetic properties, which can be easily tailored through doping, alloying, and nano engineering.

Many techniques have been employed to fabricate 1-D nano architectures, such as EBL [24], NIL [25], VLS [26], CVD [27], sol-gel [28], hydrothermal process [29], and thermal evaporation [30]. Electrochemical deposition demonstrates another important approach to the synthesis of 1-D nanostructures [31]. This approach is promising in terms of cost and ease of mass production.

Recently, the stress-related issues in the nanostructures have been considered for the integration of nano devices with high performance and functionality. Several aspects play a vital role in the control of stresses within the 1-D nanostructures [32-34]. Sheng et al. [33] demonstrated the conversion of mechanical energy into a 1.2-V electrical energy due to the 0.19% of strain induced in the aligned ZnO NWs. The electrical, optical, and magnetic properties of 1-D nanostructure are also affected by the residual stress [34]. Zhang et al. [35] and Azaceta et al. [32] reported the *c*-axis-oriented ZnO film growth and observed that thermal annealing treatment eliminates residual stress of the film. Seipel et al. [36] demonstrated that the variation in electrodeposited ZnO nanostructure crystal stresses is due to interstitial defects, voids, etc. Agrawal et al. [37] explained the importance of uni-axial stress between bulk and thin films. All the above works are not related to the effects of the nanostructure/substrate interface stresses which can influence the structural, electronic, and magnetic properties of nanowire (NW) array. There are various proposed mechanisms for intrinsic residual stress generation [38]. One possibility that has been often cited for the compressive intrinsic stress is the development of free surfaces or variation in crystal size of nanostructures before other competing stress is generated during process.

The interfacial interaction of a nanostructure with its substrate is a critical issue from the technological viewpoint, which is not clearly addressed previously. In particular, the stress-related issues in ultrathin and thin NW array grown by one-step electrochemical deposition have also not been investigated. Raman scattering has been extensively used to investigate the oxide materials, and it is a proven method for analyzing the residual stress in films [39]. However, Raman spectroscopy has not been used for stress analysis of NW arrays.

Further, the tunable ferromagnetic properties of ZnO show great potential for the use in spintronic and magneto-optic-based devices. As the dimensions of ZnO nanostructures are comparable to that of its exciton Bohr radius (approximately 2.4 nm), interesting optical emissions have been observed due to the band gap engineering from 2.5 to 6 eV by alloying. Xu et al. [40] investigated magnetic properties of ZnO nanostructured film of approximately 15-nm grain size deposited by a sol-gel technique at room temperature. However, the microscopic origin of this ferromagnetic transition is poorly understood. Moreover, the variation in crystallite size poses the serious consequences on mobility of nanostructures [41,42]. Understanding and control of these properties are the most important challenges in magnetism and photonics for the use of ZnO NWs in spintronics, lasing, and magneto-optic devices.

Recently, we reported the growth of approximately 100-nm ZnO NWs anchored on a substrate by a chronoamperometry method [43]. Here, we consider a size-dependent (approximately 15 to 100 nm) study of one-step electrochemically grown, well-aligned ultrathin and thin ZnO NW arrays before and after ultrahigh vacuum (UHV) thermal annealing treatment. We address residual stress-generated critical issues related to structural, optical, and magnetic properties. The generation of stress influences the various crystallographic axes of 1-D ZnO NW arrays, or a range of defects and complexes, causing lattice expansion or contraction. Therefore, examining the stress in the NW arrays could provide useful information on the defect evolution, which is very important to better understand and improve the film's electrical, optical, and magnetic properties. On the other hand, it is realized that the band structure of 1-D ZnO NW arrays and localized electron spin interaction may change with the stress field and thus modify the optical, electrical, and magnetic characteristics. The morphological, structural, and compressive stress analysis of 1-D NW array is carried out by field emission scanning electron microscopy (FESEM), high-resolution transmission electron microscopy (HRTEM), X-ray diffraction (XRD), and Raman spectroscopy, respectively, before and after thermal annealing treatment. Similarly, hot probe and vibrating sample magnetometer (VSM) analyses were performed to check the conductivity and magnetization properties of the 15- and 100-nm 1-D ZnO NW arrays.

METHODS

Synthesis of 1-D NW Array

A typical synthesis of 1-D well-aligned ZnO nanowire arrays was carried out using Auto Lab PGSTAT Model302 potentiostat/galvanostat (Metrohm Autolab B.V., Utrecht, The Netherlands) which involves a three-electrode cell system. Cyclic voltammetry measurements were collected at a scanning speed of 20 mV/min. All the reagents used were RA grade. The electrolyte used for ZnO NW array growth was $Zn(NO_3)_2 \cdot 6H_2O$ ((0.1 M), 2.978 g in 100 ml H_2O) dissolved in MilliQ (Millipore Corporation, Billerica, MA, USA) approximately 18 MΩ water. Copper tape served as the cathode substrate for the ultrathin and thin NW array growth. The electrodeposition was carried out at −1.2 V at 75°C; the detailed synthesis process of 1-D NW is given elsewhere [43]. After that, following rinsing with water and ethanol, these as-grown ultrathin (15 nm) and thin (100 nm) well-aligned ZnO NW array samples were subjected to thermal annealing treatment at 473, 673, and 873 K for 1 h in the presence of UHV of the order of 2.3×10^{-6} mBar. The schematic illustration of the electrochemical deposition process for the well-aligned ZnO nanowire arrays is shown in Figure 1.

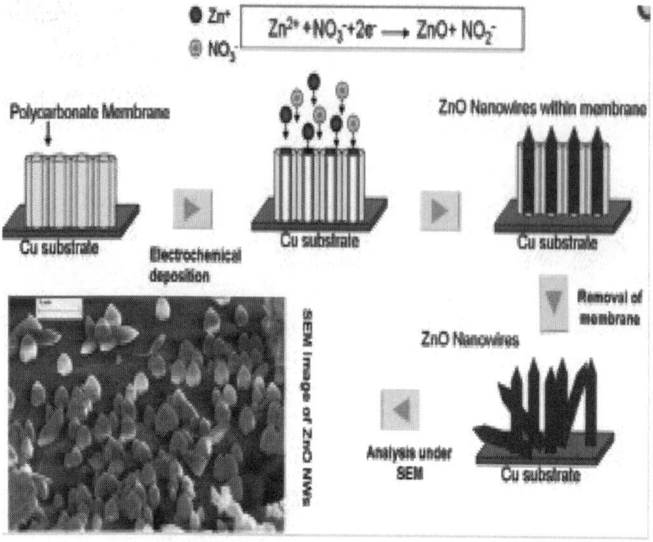

Figure 1: Template-based electrochemical deposition process for well-aligned ultrathin (15 nm) and thin (100 nm) ZnO nanowire arrays.

Characterization Techniques

The FESEM images of the grown ZnO nanowire arrays were captured using a Carl Zeiss Supra 40 microscope (Oberkochen, Germany) at an acceleration voltage of 10 kV. The hot probe analysis of ultrathin (15 nm) and thin (100 nm) ZnO NW arrays before and after thermal annealing treatment was carried out for majority charge carrier analysis; the detailed specifications are given elsewhere[43]. High-resolution transmission electron microscopy, FEI (TECNAI, Hillsboro, OR, USA), having selected-area electron diffraction (SAED) facility was used for structure analysis of the grown NW such as crystallinity, d-spacing, and c-axis orientation. The XRD experiments have been conceded in order to characterize the structural evaluation of the as-grown ultrathin and thin ZnO NW arrays before and after thermal annealing treatment. Thin film XRD spectra of the ZnO NW arrays were taken on a PANalytical X'Pert PRO diffractometer (Almelo, The Netherlands) operating in the to 2 Bragg configuration using CuK (=1.5405) radiation. Data were collected at a scan rate of 0.02° s^{-0} and a sampling interval of 0.0197°. The voltage was set at 45 kV with a 44-mA flux. The relative percentage intensities and crystalline quality of the NW array peaks were computed, considering the highest and full-width half maximum (FWHM) of [002] peak for ZnO NWs [44-48]. The stress analysis of the ZnO nanowires was carried out using a WiTec CRM 200 micro-Raman spectrometer (Ulm, Germany) coupled with a high-resolution confocal optical microscope with a laser excitation of 514.5-nm; a detailed specification is given in our previous work [43]. Transitions between weak and strong magnetic states of ZnO NW arrays before and after UHV treatment were studied using ADC Technologies Model 32KP (GMW Magnetic system; Newport Beach, CA, USA) VSM after the necessary background diamagnetic subtraction. The magnetization of the copper substrate was first measured and corrected for the experimental data. In this work, the magnetic data are presented in emu/g.

RESULTS AND DISCUSSION

In this study, the ZnO nanowires were synthesized by using the chronoamperometry method [43]. Typically, the deposition process can be delineated into three different zones as shown in Figures 2and 3. When the voltage is applied, a slight increase in the current at the beginning of region (I) of deposition indicates the filling of the pores, whereas a decrease in current indicates formation of the diffusion layer on the surface [44,45]. During the growth of the ZnO NWs in the pores, the current remains nearly constant in region (II). Once the ZnO nanowires fill the pore and approach the polymer membrane surface, caps start to grow laterally. This results in an increase in the current due to the increase in the surface area as shown in region (III). In this work, ultrathin and thin ZnO NW arrays are grown at a fixed reduction potential −1.2 V. The approximate deposition time was recorded between 75 to 100 s for ultrathin (approximately 15 nm) and 100 to 125 s for thin (approximately 100 nm) NW arrays. Thus, the concerted effect of deposition potential, diffusion of reacting species, and the effect of pore diameter with pore filling rate govern the aspect ratio of the grown ZnO nanowire arrays embedded in polycarbonate template as described in our previous work[43].

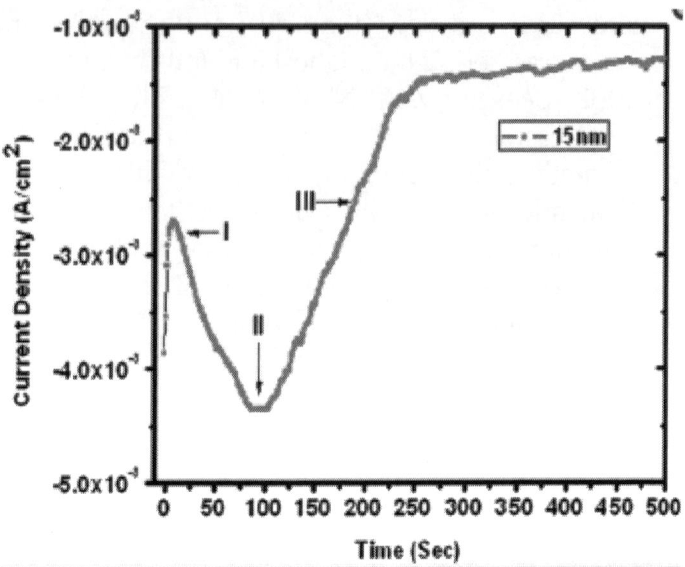

Figure 2: Current-time curve of ultrathin (15 nm) ZnO nanowire arrays grown electrochemically within pores of polycarbonate membrane.

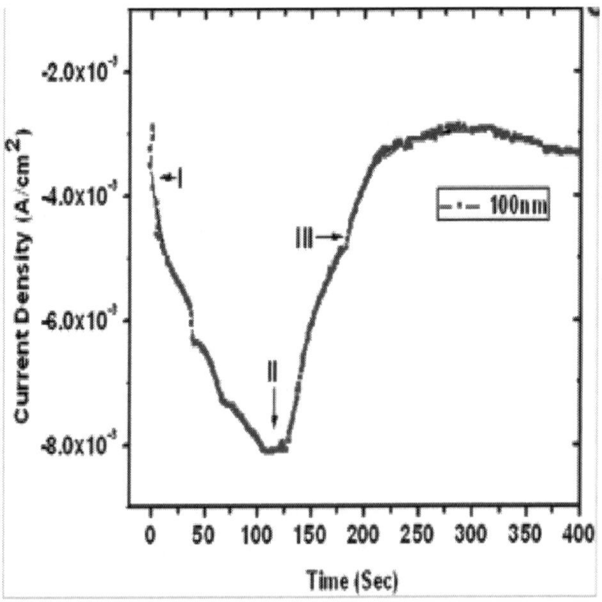

Figure 3: Current-time curve of thin (100 nm) ZnO nanowire arrays grown electrochemically within pores of polycarbonate membrane.

The SEM images in Figures 4 and 5 show the aspect ratios of the ultrathin (approximately 15-nm diameter) and thin (approximately 100-nm diameter) ZnO NW arrays to be approximately 133 and 20, respectively. In what follows, we will consider these two diameters of NW arrays, thin and ultrathin, for comparison and contrast with regard to their properties. The XRD patterns for the as-grown ultrathin and thin NW arrays are shown in Figure 6 peak a and Figure 7 peak a, respectively. The XRD results after UHV thermal annealing treatment at temperatures of 473, 673, and 873 K are shown in Figure 6 peaks b, c, and d) for the ultrathin aligned NWs; the results are the same for the thin wire arrays in Figure 7 peaks b, c, and d. The XRD diffraction of the electrochemically grown ZnO nanowires confirmed a polycrystalline wurtzite crystal structure with lattice parameters of $a = 3.24$ Å and $c = 5.20$ Å. From thermodynamic consideration, the *c*-axis is the preferred orientation of the grown NWs where the crystallites perpendicular to Cu substrate can be expected [46,47]. The crystal structure and crystallites demonstrate an elongation in the [002] direction and a contraction in the [100] direction with respect to bulk films [46]. The intensity of ZnO [101] reflection remains maximum for all samples, while the intensities of [100] and [002] reflection planes vary with the thermal annealing treatment. The computed relative intensities for the as-grown NWs and after UHV thermal annealing treatment are tabulated in Tables 1 and 2.

Figure 4: FESEM images of well-aligned ZnO nanowire arrays (approximately 15 nm) grown on Cu substrate. (a) Low-magnification image from top view.

(b) Aligned nanowires at a viewing angle. (c) Top view at one point. (d) Another view at another point. (e) High-magnification image of vertically aligned nanowires. (f) Dimensional measurements of a single nanowire.

Figure 5: FESEM images of approximately 100-nm ZnO nanowire arrays. (a) Low-magnification image of vertically aligned nanowires. (b)High-magnification image of nanowires. (c) Inset view of the aligned nanowires. (d) Low-magnification image at a different point. (e) High-magnification image of the same. (f) Hexagonal structural growth of ZnO nanowires. (g) Measured dimensions of nanowires.

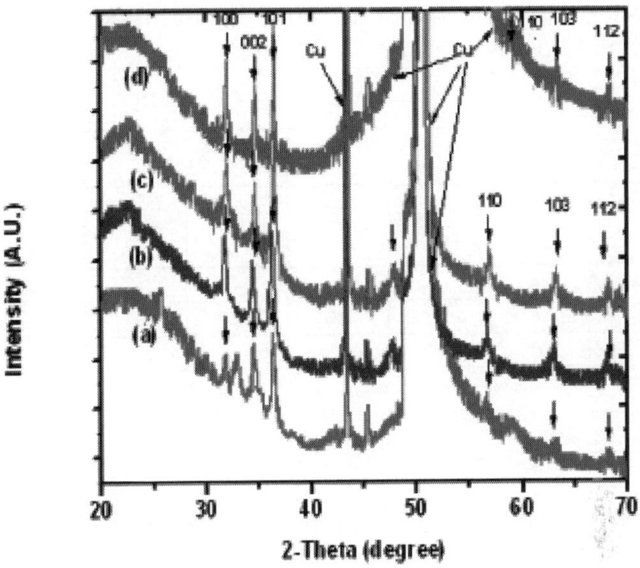

Figure 6: XRD diffractogram of electrochemically grown ZnO nanowire arrays at approximately 15 nm. (a) XRD of as-grown (300 K). (b) UHV-annealed at 473 K. (c) UHV-annealed at 673 K. (d) UHV annealed at 873 K.

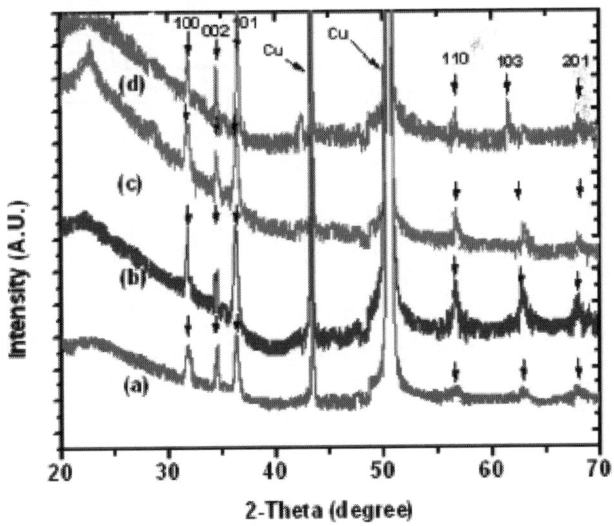

Figure 7: XRD diffractogram of electrochemically grown approximately 100-nm ZnO nanowire arrays. (a) As grown (300 K). (b) UHV-annealed at 473 K. (c) UHV-annealed at 673 K. (d) UHV-annealed at 873 K.

Table 1: Variation in peak intensities measured from XRD diffractograms of ultrathin (15 nm) ZnO NW arrays

ZnO peak	300 K (room temperature)		473 K		673 K		873 K	
	Intensity	Relative intensity	Intensity	Relative intensity	Intensity	Relative intensity	Intensity	Relative intensity
100	190.0	0.24	254.1	0.39	264.9	0.4	279.2	0.47
002	211.5	0.26	254.0	0.45	346.9	0.63	357.9	0.68
101	342.5	0.64	451.1	0.72	530.9	0.85	569.9	1.22

Shrama *et al.*

Shrama *et al. Nanoscale Research Letters* 2014 9:122 doi:10.1186/1556-276X-9-122

Table 2: Variation in peak intensities measured from XRD diffractograms of thin (100 nm) ZnO NW arrays

Variation in peak intensities measured from XRD diffractograms of thin (100 nm) ZnO NW arrays								
ZnO peak	300 K (room temperature)		473 K		673 K		873 K	
	Intensity	Relative intensity	Intensity	Relative intensity	Intensity	Relative intensity	Intensity	Relative intensity
100	114.7	0.02	155.1	0.03	257.4	0.1	312.1	0.13
002	119.4	0.03	162.7	0.06	248.8	0.07	261.8	0.11
101	264	0.03	389.8	0.06	460.0	0.25	623.6	0.26

Shrama *et al.*

Shrama *et al. Nanoscale Research Letters* 2014 9:122 doi:10.1186/1556-276X-9-122

The percentage of relative intensity of the [002] peak varies from 0.23 to 0.68 cts for as-grown to the thermally treated ultrathin NW arrays at 873 K. Similarly, this value varies from 0.03 to 0.11 cts for thin NWs. Thus, thermal treatment of NW arrays at 873 K causes a significant shifts of approximately 0.45 and approximately 0.08 cts for ultrathin and thin NWs. Thus, annealing effect enhanced the fraction of crystallites in well-aligned ZnO NW arrays [48]. These XRD studies confirm that the as-grown 15-nm ZnO NWs are found to have higher crystallinity than 100-nm NWs.

After UHV treatment, this effect is observed to be more pronounced (by a factor of 5) for ultrathin NWs as compared to that for thin NW arrays. It indicates that 15-nm ZnO NWs are much better suited for optical emission-based applications than 100-nm NWs. Furthermore, this increase in intensities also indicates higher mobility of their atoms for the ultrathin than the thin NW array. It may be attributed to the presence of higher number of defect densities at the surface of ultrathin NWs. Meng et al. [49] also reported similar behavior for thin films of zinc oxides. In order to estimate the variation in crystallite size (*D*) of ZnO NWs before and after thermal treatment, the FWHM of [002] diffraction peak is used in the Scherrer equation [50]:

$$D = \frac{0.9\lambda}{B\cos\ \theta}, \tag{1}$$

where , , and *B* are the X-ray wavelength, Bragg diffraction angle, and FWHM, respectively. There is a gradual increase observed in the crystallite size of the 15-nm NW array as compared to that of the 100-nm NW array as shown in Table 3. A significant shift in crystallite size of approximately 3.3 nm and approximately 24.7 nm was observed by thermal annealing at 873 K for the ultrathin and thin ZnO NW arrays, respectively. Similar observations have been made in the context of thin films[47-50]. Apart from a shift in crystallite size, another essential issue of ultrathin and thin ZnO NW arrays is the variation in residual stress before and after UHV thermal annealing treatment. The origin of stress or strain generation is the lattice mismatch, due to difference in the thermal expansion coefficient and internal stress related to the defect of nanostructures and substrate [51]. It has been realized that the electrical, optical, and magnetic properties may be altered by varying the stress or strain of the films/nanostructures, which can also be altered by thermal annealing. The strain in thec-axis could be calculated and expressed by the following formula [46]:

$$\varepsilon = \frac{C - C_0}{C_0}, \tag{2}$$

where C and C_0 (0.5205 nm) are the strained and unstrained lattice constant, respectively. The lattice constant could be computed according to the Bragg formula and peak position in the XRD pattern. The stress in the well-aligned ZnO NWs is computed by the following relation [50]:

$$\sigma = -453.6 \times 10^9 \left(\frac{C - C_0}{C_0} \right) \tag{3}$$

Table 3: FWHM, d-spacing, crystallite size, and compressive stress variation computed from the XRD diffractograms

15-nm ZnO NWs				100-nm ZnO NWs				
Temperature (Kelvin)	FWHM (degrees)	d-spacing ( )	Crystallite size (nm)	Stress (GPa)	FWHM (degrees)	d-spacing ( )	Crystallite size (nm)	Stress (GPa)
300	0.741	2.6012	12.8	0.2248	0.4315	2.6017	22.8	0.1359
473	0.684	2.5974	13.2	0.8888	0.3313	2.5978	30.2	0.8190
673	0.593	2.5963	15.5	1.0806	0.2880	2.5965	35.7	1.0457
873	0.572	2.5954	16.1	1.3072	0.2300	2.5958	47.5	1.1677

Shrama *et al.*

Shrama *et al.Nanoscale Research Letters* 2014 9:122 doi:10.1186/1556-276X-9-122

The stress computed from the above relation for the ultrathin and thin ZnO NW arrays before and after thermal annealing treatment is shown in Table 3. Similarly, Figures 8 and 9 show the variation of compressive stress as a function of temperature for ZnO NW arrays. The shift in magnitude of compressive stress for as-grown to the UHV thermally annealed at 873 K of ultrathin and thin ZnO NW arrays is approximately 1.08 GPa and approximately 1.03 GPa, respectively. The positive magnitude of stress clearly indicates that the grown NW array is in a state of compressive stress instead of tensile stress [50,51]. These results show that the ultrathin ZnO NW arrays are at higher compressive stress than the thin ZnO NW arrays. However, the ZnO nanowires are *c*-axis oriented, which suggests that the compressive stress is also preferentially oriented along the same direction. One can conclude that the significant shifts in compressive stress could be related to the disparity between the thermal expansion coefficient of ZnO NW array with the Cu substrate. The thermal expansion coefficient ($_{ns}$) of ZnO NW array is 2.9×10^{-6} K^{-1}, while thermal expansion coefficient ($_{s}$) for Cu substrate is 16.4×10^{-6} K^{-1}. Similarly, Fang et al. [50] demonstrated the increase in compressive stress for a ZnO film after annealing treatment. Another possible factor for the generation of compressive stress in ZnO NWs is the presence of zinc defects in the interstitial voids, causing disordering of the lattice. Also, atoms on a free surface have missing bonds and therefore tend to reorganize by the UHV thermal annealing to maximize bonding. Similar stress-induced phase transformation in ZnO NWs has been previously reported by Kulkarni et al. [52]. Lee et al. [53] reported the reduction in internal stress of ZnO film from compressive to tensile by variation of temperature. However, the grown ultrathin and thin ZnO NWs are polycrystalline and have a very strong preferred orientation of their crystallites in the [000] direction perpendicular to the copper substrate as shown in Figure 10a,b. The thermal expansion coefficient for Cu is approximately eightfold higher than those of the ZnO NW arrays. Thus, the lattice mismatch between the ZnO NWs and Cu substrate may be the deciding factor for the stress generation at the interface. It is reported that the presence of higher thermal expansion coefficient difference between the substrate and grown nanostructures often gives a higher compressive stress [50]. Compressive stress would be the driving force responsible for the orientation of the NW crystal faces. According to the 'supercell' approach, ZnO tetrahedral nanostructure terminates with the four most important growth surfaces, the non-polar $[10\overline{1}0]$ and [1120] and the polar [0001] and $[000\overline{1}]$ as shown in Figure 11a,b and Figure 10b. Zn^{2+} (0001) and O^{2-} $[000\overline{1}]$ surfaces are of considerable interest because of their polar nature [54]. The polar character of [0001] surface is the requirement for the surface reconstructions. On the other hand, the surface of Cu substrate (fcc structure) facet orientations are likely to be favored in the [111] plane and [110] plane. Therefore, the effective residual stress in the Cu substrate could be because of the most preferential plane [111]. In this way, the Cu substrate may offer the adsorption/desorption sites for ZnO NW array at the interface, bridging sites and hollow sites, as shown in Figure 12a,b. The ZnO $[000\overline{1}]$-O and ZnO [0001]-Zn surfaces are terminated in a hexagonal array of nearly closed packed oxygen anions and zinc cations as shown in Figure 12a,b. This rearrangement of NW substrate interface is probably due to the thermal vacuum annealing that restructures the facets or recystallizes new crystal faces other than [0001] and $[10\overline{1}0]$. Additionally, the diffusion of oxygen atoms from ZnO $[000\overline{1}]$-O could be formed by the bonding with the copper interface, which leads to the generation of increase in compressive stress due to the annealing effect. Generally, Cu metal substrate is electron rich and therefore naturally repels the oxide anions while being attracted toward the Zn cations as shown in Figure 12a. Interaction between oxygen ions at the surface of ZnO close to the Cu anchor has been reported recently [54,55]. This interaction could be understood as a result of polarization of the oxide anions while the Cu^{+} ions are located in the surface plane as shown in Figure 12b.

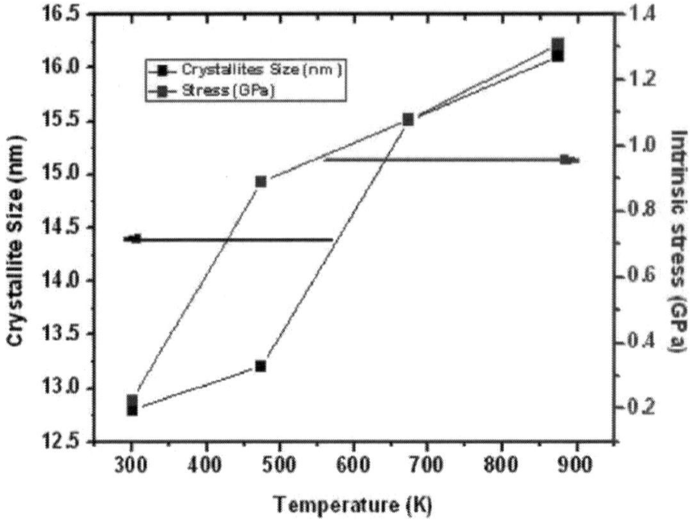

Figure 8: Dependence of the crystallite size and stress variation for ultrathin (15 nm) ZnO NW arrays.

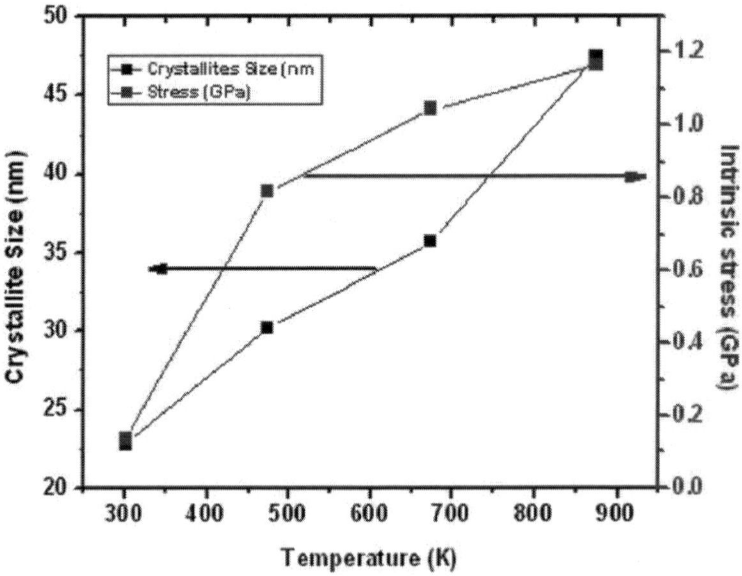

Figure 9: Dependence of the crystallite size and stress variation for thin (100 nm) ZnO NW arrays.

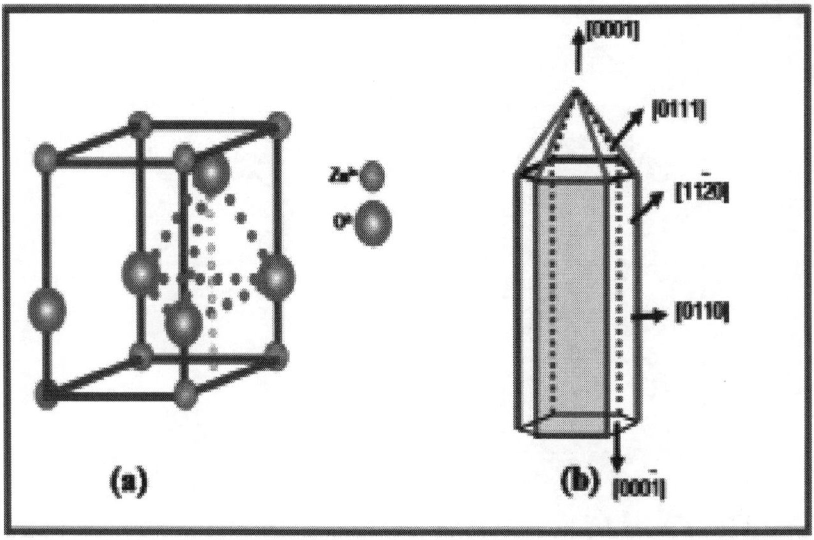

Figure 10: Atomic arrangement of wurtzite structures and the structure of ZnO NWs. (a) Atomic arrangement in crystal structure of tetrahedral wurtzite structures of ZnO. (b) Structure of ZnO NWs with different planes.

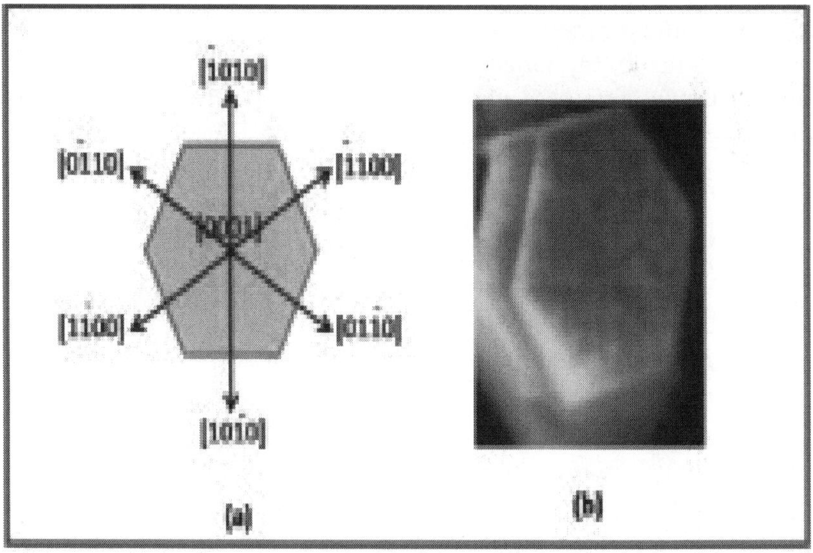

Figure 11: Hexagonal structure and SEM image of ZnO nanowire. (a)Directions of the panes in the hexagonal structure. (b) SEM image of the top view of a single grown ZnO nanowire.

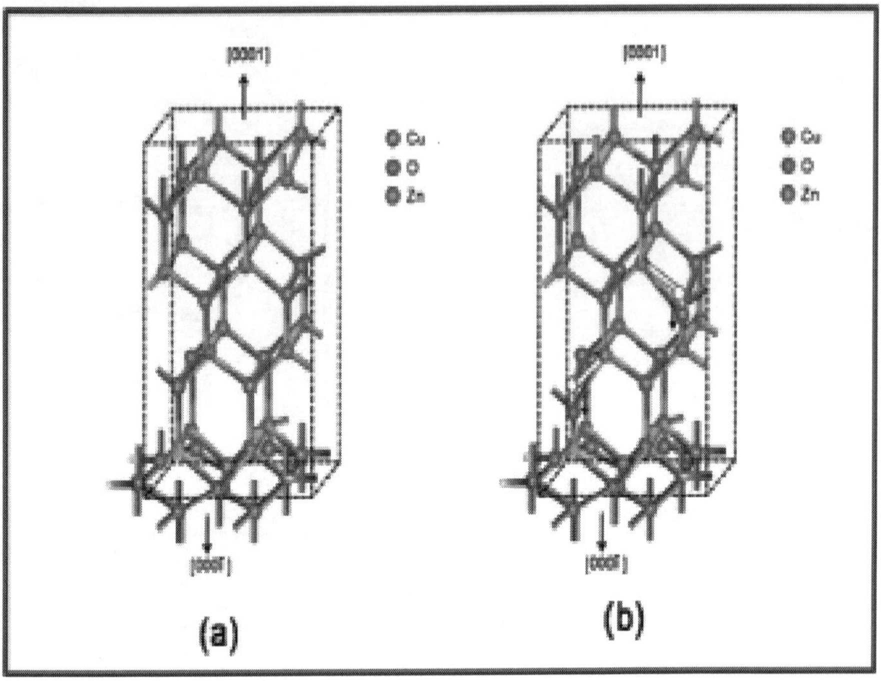

Figure 12: Atomic arrangement, interface interaction, and O2diffusion mechanism. (a) Atomic arrangement within ZnO NWs and interface interaction with Cu substrate. (b) O_2 diffusion mechanism to the Cu substrate due to shift in bond alignment and reconstruction of bonds after UHV thermal annealing treatment.

Likewise, the residual compressive stress drives the Zn ions of the outermost layer inward and results in the reduction of Zn-O bond length of dimmers due to thermal annealing. The surface energies of the polar surfaces [0001] and $[100\bar{1}]$ are greater than those of the non-polar $[10\bar{1}0]$ and [1120] surfaces for all nanostructures [53-55]. The variations in the bond length of Zn-O dimmers cause tilting from its alignment perpendicular to the direction as shown in Figure 12b. The variation in compressive stress related to the change in bond length alignment in the dimmers can cause the increase in Zn concentration as a result of adsorption/desorption of oxygen atoms within NW array due to the effect of thermal annealing treatment. The formation of newly crystal faces as a consequence in the variation of crystallites and grain boundaries related to the compressive stress could be another, possible reason. It is thus concluded that lattice mismatch, defects, grain boundaries, NW/substrate interaction, and annealing treatment are major causes for the stress variations.

In order to elucidate the crystal structures of the ultrathin (15 nm) ZnO NW, low- and high-resolution (HRTEM) analyses, together with SAED patterns, are investigated; the results are shown in Figure 13a,b,c,d. The HRTEM images of the 15-nm ZnO NWs are shown in Figure 13a,b, which clearly shows that the ultrathin ZnO NWs have a diameter of around 15 nm and have a high aspect ratio (132 to 200 nm). This significant variation in the aspect ratio in HRTEM images might be due the polydispersity and fragile nature of the one-dimensional ultrathin (15 nm) ZnO NW arrays. This consequence supports the FESEM outcomes shown in Figure 4, instigated from one-step template-assisted electrodeposition process for ultrathin ZnO NW arrays. Furthermore, the HRTEM images of Figure 13a,b evidently show that the ultrathin ZnO NWs are uniform, straight, have sharp tips, with diameters of approximately 15 nm, in the [0001] orientation, and have high aspect ratio (approximately 132 to 200 nm) as demonstrated in the FESEM images of Figure 4. Similar approach results shown for the thin ZnO NW arrays are not shown here. Figure 13c,d depicted the lattice spacing and SAED pattern of ultrathin ZnO NW arrays. The lattice spacing of 0.58 nm was estimated from the corresponding d-spacing of the adjacent planes ±0001, as shown in Figure 13c. The SAED bright spots indicate ultrathin ZnO NWs, crystal planes of the hexagonal structure, polycrystalline nature, and *c*-axis orientation. These results support the previous XRD investigation about the polycrystalline nature of ultrathin ZnO NW arrays. Likewise, thin ZnO NW array also shows the polycrystalline nature; results are not shown here. Among the many similar characteristics, the foremost disparity found in thin ZnO NWs as compared to the ultrathin NWs was their multi-grains boundary.

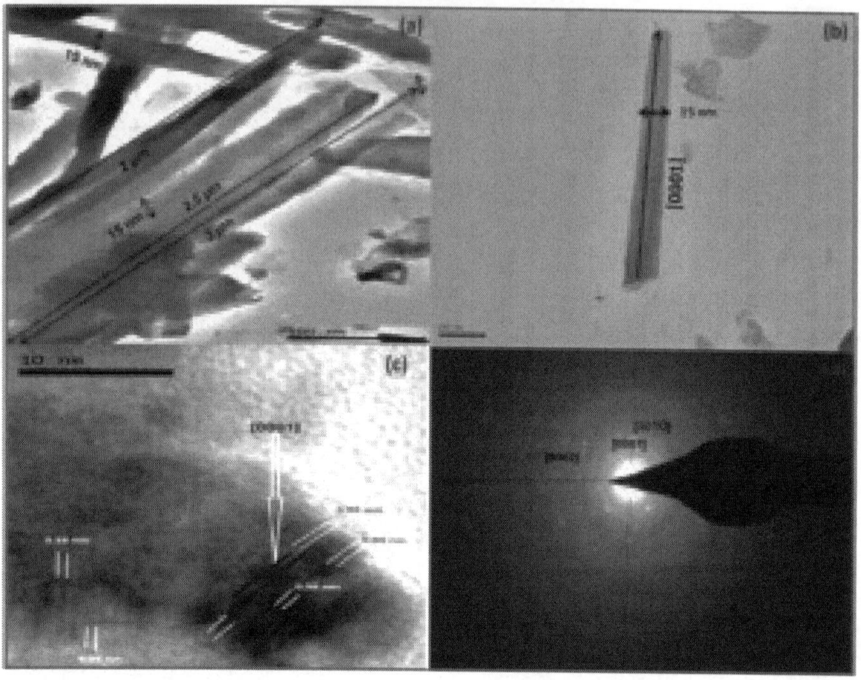

Figure 13: HRTEM images of ultrathin (15 nm) ZnO nanowire array. (a) Low-magnification image of ZnO NWs synthesized by electrochemical deposition. (b) High-resolution TEM image of 15-nm diameter and [0001] crystal plane. (c) TEM image of lattice fringes suggesting the *c*-axis growth direction and spacing of 0.58 nm in the adjacent lattice planes. (d) SAED diffraction pattern indicating the polycrystalline nature of grown ultrathin ZnO NW array.

In the wurtzite-type ZnO, the lattice symmetry is mainly reflected by two types of lattice phonons $A_1$and E_1. Both A_1 and E_1 symmetries are Raman active, and each splits into longitudinal optical (LO) and transverse optical (TO) components with different frequencies due to the macroscopic electronic fields associated with the phonons [56]. The Raman spectra for well-aligned as-grown ultrathin and thin NW arrays and for NW arrays after UHV treatment are shown in Figures 14 and 15. The estimated values of E_1(LO) mode of FWHM for the as-grown ultrathin and thin ZnO NW arrays before and after ultrahigh vacuum thermal annealing treatment at 473, 673, and 873 K are 16.6, 23.0, 25.6, and 31.3, and 27.4, 28.3, 30.1, 34.7, respectively. It evidenced a considerable shift approximately 14.7 and approximately 7.3 in FWHM values for the as-grown ultrathin and thin NW arrays thermally treated at 873 K as shown in Figures 14 and 15. This marked shift indicates the presence of higher oxygen vacancies in 15-nm NW arrays than in the 100-nm ZnO NW arrays. Additionally, the enhancement of FWHM value after UHV thermal annealing treatment corresponds to the redistribution of intrinsic defect of ZnO NW arrays. Moreover, the E_2 (high) vibration mode phonon frequency reflects the induced stress in wurtzite ZnO nanostructures [57]. Experimentally, E_2 (high) and E_2(low) phonon modes are observed at approximately 432 cm^{-1} and approximately 96 cm^{-1}. The E_2(high) vibration mode frequencies for as-grown ultrathin and thin ZnO NW arrays estimated before and after thermal annealing treatment are 433, 435, 439, and 444 rel cm^{-1}, and 432, 434, 437, and 440 rel cm^{-1}, respectively, as shown in Figures 14 and 15. This increase in E_2 (high) phonon frequency ascribed to the higher compressive stress in 15-nm NWs instead in 100-nm ZnO NW arrays. Besides this, the quantum confinement effect is more significant in ultrathin NW array as compared to thin NW array. Additionally, as shown in Figure 16, there is a significant shift in magnitude of FWHM E_2 (high) phonon vibration modes of approximately 8.6 and approximately 25.8 for the as-grown ultrathin and thin ZnO NWs compared to the ultrahigh vacuum annealing treatment at 873 K. This noticeable FWHM shift by a factor of 3 for thin NWs with respect to high-aspect-ratio ultrathin NWs attributes to the wider size structural redistribution after UHV treatment. On the other hand, the increase in E_2 peak intensity with the increase of annealing temperature was also observed. Therefore, the decrease in FWHM and the increase in peak intensity both indicate the increase in crystallite size and an improvement in the crystalline quality of well-aligned ZnO NW arrays after UHV annealing. These results demonstrate that 15-nm ZnO NWs have lower lasing power threshold than 100-nm NWs due to the higher crystallinity of the ultrathin ZnO NW arrays. This effect shows significant improvement after UHV annealing. These results support our XRD investigation of variation in the crystallite size and compressive stress for well-aligned ultrathin and thin ZnO NW arrays, before and after thermal annealing treatment. Similarly, the broader peak at 328 cm^{-1} is attributed to multiple-phonon scattering processes. An additional peak at 206 cm^{-1} is observed, and its origin is still not clear [58]. The A_1(LO) and E_1(LO) peaks at approximately 542 and 580 cm^{-1}, respectively, are related to smaller scattering cross-sections [37,58]. The Raman peak that reveals the two-phonon combination,

A_1(LO) + E_2 (high) scattering, is noticed at 980 rel cm^{-1}[37], while the peaks at 1,360 and 1,460 rel cm^{-1} represent the phonon scatterings A_1(TO) + A_1(LO) + E_2 (high) and A_1(LO) + E_1(TO) + E_2 (high). Increase in intensity from 980 to 2,000 rel cm^{-1} as the effect of UHV annealing treatment from 473 to 873 K is a result of multiple-phonon scattering, due to the increase in crystallite size [37,57,58]. It is probably because of the ultrahigh vacuum annealing which coalesces the small crystallite.

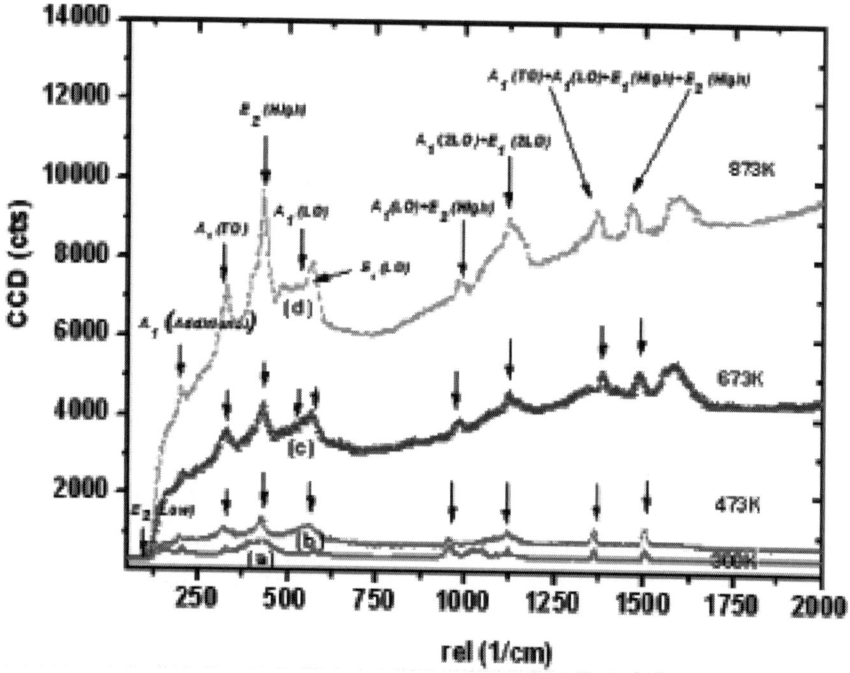

Figure 14: Micro-Raman spectra of approximately 15-nm (ultrathin) ZnO nanowire array. (a) As-grown (300 K). (b) UHV-annealed at 473 K. (c) UHV-annealed at 673 K. (d) UHV-annealed at 873 K.

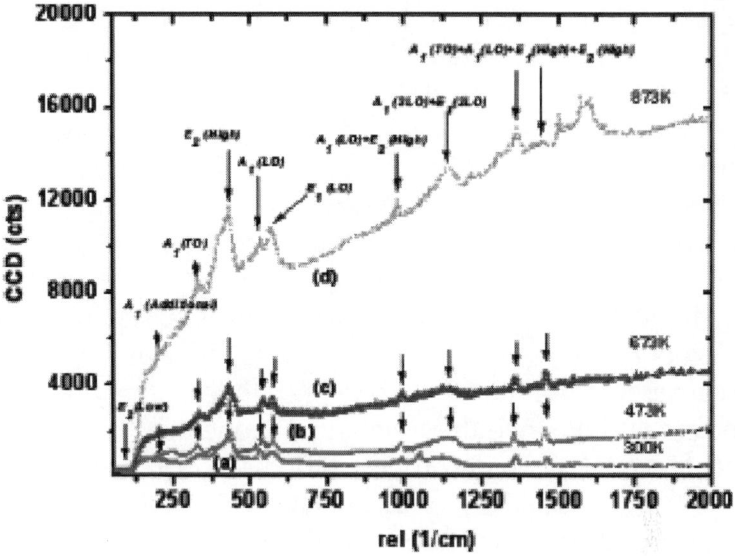

Figure 15: Raman spectra of approximately 100-nm (thin) ZnO nanowire arrays. (a) As-grown (300 K). (b) UHV-annealed at 473 K. (c) UHV-annealed at 673 K. (d) UHV-annealed at 873 K.

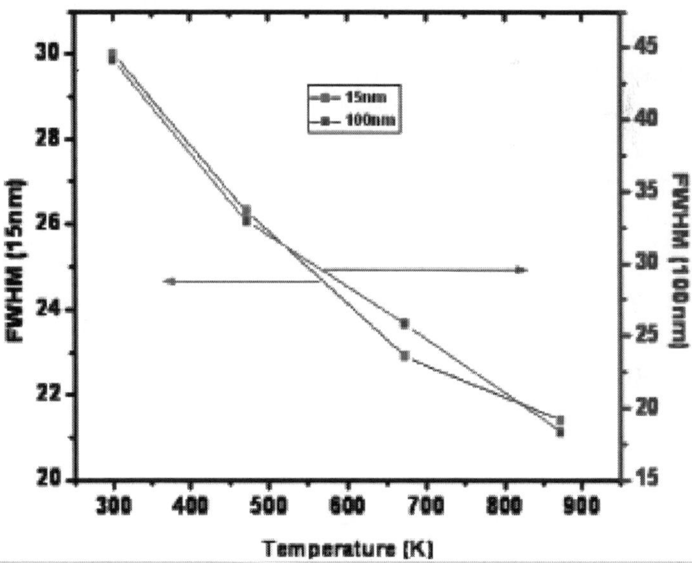

Figure 16: Variation in FWHM computed from Raman spectra plotted as a function of annealing temperature.

The estimation of majority charge carriers in the ZnO NW array is done through hot probe technique. The ultrathin and thin as-grown and thermally treated NWs at 473, 673, and 873 K in the presence of UHV show positive voltages of 320, 360, 392, and 482 mV, respectively, for 15-nm NW arrays and 109, 157, 205, and 286 mV, respectively, for 100-nm NW arrays. The positive magnitude of voltage indicates that the grown ZnO nanowires are n-type semiconductors. The higher values of readout voltage for ultrathin as compared to the thin indicate that the ZnO segregation as a separate phase in the grain boundary regions results in the higher number of majority charge carries for ultrathin NW arrays [42]. Indeed, the most important parameters that influence their electrical properties are the crystallite size and the adsorption/desorption processes of oxygen. Moreover, the changes in conductivity of ZnO NW arrays with UHV annealing most likely result from variations in the number of oxygen species adsorbed/desorbed on the ZnO surfaces or in the number of oxygen vacancies in the ZnO bulk. Therefore, the UHV annealing in vacuum should decrease the number of adsorbed oxygen species onto the surface and increase the number of bulk oxygen vacancies; these act as electron donors for ZnO NW arrays. This effect is more prominent for the ultrathin as compared to the thin NW arrays because of the small dimensional quantum confinement and higher oxygen vacancies in the former. It might have the order of 10^{20} surface oxygen sites per cubic centimeter of nanostructures. Thus, even for partial changes in the concentration of adsorbed oxygen species, large changes in NW arrays conductivity can be observed. Hence, the strong dependence of the conductance on the oxygen vacancies occupancy in ZnO 1-D NW arrays is an important characteristic of functional oxide, using one that can tune and control the electrical properties of the nano device, especially the threshold voltage of ZnO-based field effect transistors (FinFETs and MOSFETs). Nevertheless, the charge carrier behavior may be a composite effect of oxygen vacancy, oxygen interstitial site, zinc vacancy, and zinc interstitial site within the ZnO NW arrays. Therefore, the evolution of the majority of charge carrier mobility does not follow an expression rigorously proportional to crystallite size and oxygen vacancy [41,42,56,59]. As shown in Table 3, the gradual increase in the crystallite size for all samples with the annealing effect of UHV treatment reveals the redistribution of oxygen vacancies and deoxidization of trap density. Similar evolution was also observed by

Kishimoto et al. [60] in undoped ZnO thin films for the critical film thickness of approximately 130 nm.

Materials with important combinations of properties such as room-temperature ferromagnetism and semi-conductive properties are required for spintronic and magneto-optic device application. Novel functionalities can be achieved, for example, in spin-FETs or spin-LEDs, if the injection, transfer, and detection of carrier spin can be controlled electrically or optically. ZnO-based ultrathin and thin NWs are thought to be ideal systems for spintronics and magneto-optic device application because of two most promising material criteria: (i) ferromagnetism should be retained up to room temperature and above the room temperature and (ii) the electrical and optical properties of ferromagnetic semiconductors should allow for spin manipulation.

Figure 17 curves a, b, c, and d and Figure 18 curves a, b, c, and d show the magnetization versus applied magnetic field for the as-grown and thermally annealed samples at 300, 473, 673, and 873 K, respectively, for ultrathin and thin well-aligned ZnO NW arrays. One can see that the as-grown NW arrays are ferromagnetic, not only at room temperature (300 K) but also after annealing at high temperatures [40,61]. The measured minimum and maximum magnetizations for the as-grown ultrathin and thin NW arrays are approximately 8.45×10^{-3} to 8.10×10^{-3} emu/g and approximately 2.22×10^{-7} to 2.190×10^{-7} emu/g, respectively, as shown in Figure 17 curve a and Figure 18 curve a. Figure 17 curve a, for the as-grown ultrathin sample at 300 K, show an s-shaped behavior with magnetization saturation similar to that of a superparamagnetic material. There is no saturation of magnetization observed in Figure 18 curve a for thin NW arrays. Likewise, for the ultrathin and thin NW arrays, respectively, the computed magnitude after ultrahigh vacuum thermal annealing treatment at 473 K varies from approximately 6.08×10^{-3} to 6.26×10^{-3} emu/g and approximately 1.24×10^{-7} to 1.46×10^{-7} emu/g. At 673 K, it is approximately 5.63×10^{-3} to 5.22×10^{-3} emu/g and 9.19×10^{-8} to 8.7×10^{-8} emu/g; similarly, at 873 K, it is approximately 4.92×10^{-3} to 4.44×10^{-3} emu/g and 1.47×10^{-8} to 3.95×10^{-9} emu/g.

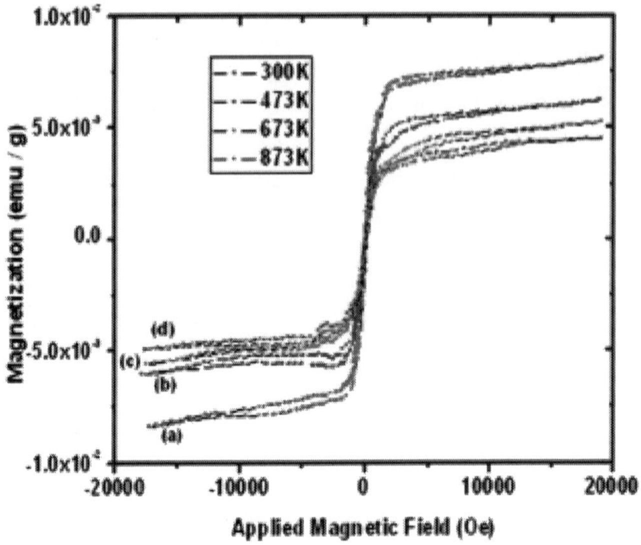

Figure 17: Magnetization response for approximately 15-nm (ultrathin) ZnO NW arrays. (a) As-grown (300 K). (b) UHV-annealed at 473 K. (c) UHV-annealed at 673 K. (d) UHV-annealed at 873 K.

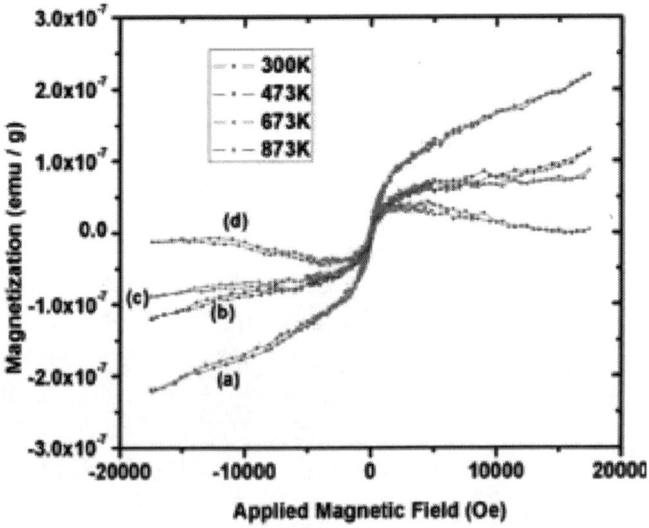

Figure 18: Magnetization response for approximately 100 nm (thin) ZnO NW arrays. (a) As-grown (300 K). (b) UHV-annealed at 473 K. (c) UHV-annealed at 673 K. (d) UHV-annealed at 873 K.

However, Figure 18 shows the hysteretic behavior and non-saturation of the magnetization for thin NW arrays even after the UHV treatment. Further, there is a steady decrease of magnetization in 15-nm as well as in 100-nm ZnO NW arrays with the UHV annealing treatment at different temperatures as shown in the Figure 17 curves b, c, and d and Figure 18 curves b, c, and d. In contrast, the saturation of magnetization for ultrathin NWs occurs at relatively lower fields, approximately 5,000 Oe. These results indicate a relatively small factor of 2 shifts in the magnetization of ultrathin (15 nm) ZnO NW arrays even after UHV annealing. In contrast, a much larger change of 2 orders is observed for thin (100 nm) NW arrays. In fact, the as-grown ultrathin NWs have higher magnetization, approximately of the order of 4 as compared to the thin ZnO NW array, though the observed magnetism in well-aligned ZnO NW arrays is unexpected, because neither Zn^{2+} nor O^{2-} is magnetic. Thus, there is no apparent source for magnetism in undoped ZnO [62]. As no magnetic impurities were present, it appears that the origin of ultrathin and thin NW array ferromagnetism may be the exchange interactions between localized electron spin moments resulting from oxygen vacancies at the surfaces of the ZnO NWs. Furthermore, the observed magnetization in the ZnO NW array may be probably due to the defects. We presumed that those defects are located close to each other and mostly are at the NW surfaces (skin effect), and after UHV treatment, reduction in magnetism indicates the decrease in the number of density defects as those in ultrathin films [63]. It supports our previous observation of XRD and Raman analyses. These studies will pave the way for the use of high-aspect-ratio 15-nm ZnO NWs as nanoscale spin-based devices, such as spin valves and spin FETs. The 100-nm ZnO NWs for deep UV magneto-optic device application with the ultimate goal of manipulating a single electron spin and polarization reflected intensity rather than the charge and source as in more conventional devices. The summary of most important scientific results of ultra-thin and thin NW array and respective device applications is in

CONCLUSIONS

We synthesized 1-D well-aligned ultrathin (15 nm) and thin (100 nm) ZnO NW arrays by the one-step chronoamperometry at reduction potential of −1.2 V. The structural, optical, electrical, and magnetic analyses of ultrathin and thin NWs are conceded by FESEM, HRTEM, X-ray diffraction, micro-Raman, hot probe, and VSM. FESEM images illustrate the aspect ratio of 133 and 20, respectively, for well-aligned 15- and 100-nm ZnO NW arrays. HRTEM and SAED patterns confirmed the polycrystalline nature of the ultrathin ZnO NWs and lattice spacing of 0.58 nm. X-ray diffraction results show the wurtzite structure of the as-grown polycrystalline ZnO NW and [002] elongation. There is higher noteworthy shift in the [002] peak intensity for ultrathin than that for thin as-grown and thermally treated NW arrays at 873 K, revealing that the 15-nm ZnO NWs are much better suited for optical emission-based applications than 100-nm NWs. Furthermore, structural stress-related critical issues on the understanding of 1-D ZnO NW arrays will provide useful information on the defect evolution, which is very important to better understand and improve the electrical, optical, and magnetic properties of nanostructures.

Therefore, the higher shift in magnitude of compressive stress for the as-grown ultrathin than thin ZnO NW arrays and UHV annealed at 873 K indicates that the 15-nm ZnO NW arrays are at higher compressive stress than the 100 nm. Micro-Raman results show the increase in E_2 (high) peak intensity, and the decrease in FWHM represents the increase in crystallite size and an improvement in the crystalline quality of NW arrays after the annealing treatment. These results demonstrate that 15-nm ZnO NW have lower lasing power threshold than 100-nm NW due to the higher crystallinity of the ultrathin ZnO NW arrays. The positive voltage for hot probe measurements points out that the grown ZnO NWs are n-type, and the higher values of voltage and vacancies for 15 nm as compared to those of the 100-nm NWs indicate the higher number of majority charge carries for ultrathin NW arrays than that for thin. Therefore, by tuning the oxygen vacancies occupancy, one can control the electrical properties of the nano device, especially the threshold voltage of ZnO-based field effect transistors (FinFETs and MOSFETs). The VSM results reveal that the as-grown 15-nm NWs have the higher magnetization approximately of the order of 4 as compared to the 100-nm ZnO NW array. In fact, there is a shift in magnetization and a propensity of saturation of magnetization for 15-nm ZnO NW arrays by factor 2 and at approximately 5,000 Oe even after UHV annealing at 873 K. In contrast, a much larger change of 2 orders and a tendency of saturation of magnetization occur at approximately 10,000

Oe for 100-nm NW arrays. These studies will pave way for the use of high-aspect-ratio 15-nm ZnO NWs as nanoscale spin-based devices, such as spin valves and spin FETs and 100-nm ZnO NWs for deep UV magneto-optic device application.

AUTHORS' CONTRIBUTIONS

The work presented here was performed in collaboration of all authors. SKS and NS carried out the synthesis and characterization measurements of electrochemical synthesis and FESEM, stress, etc. SB, RK, and SKS performed the HRTEM, XRD. AS provided guidance, supervised the work, and finalized the manuscript. All authors read and approved the final manuscript.

ACKNOWLEDGEMENTS

Authors acknowledge the support of the Department of Science and Technology (DST), New Delhi, through its financial support from DST-IRPHA grant and the Thematic Unit of Excellence (Nanoscience and Nanotechnology in Nanofabrication: Top-down, Bottom-up and Beyond) at IIT Kanpur, for providing the state-of-the-art experimental facility available at Indian Institute of Technology Kanpur.

REFERENCES

1. Lin Wang Z: ZnO nanowire and nanobelt platform for nanotechnology.*Mater Sci Eng R* 2009, 64:33-71.
2. Tian BZ, Zheng XL, Kempa TJ, Fang Y, Yu NF, Yu GH, Huang JL, Lieber CM: Coaxial silicon nanowires as solar cells and nanoelectronic power sources.*Nature* 2007, 449:885-890.
3. Hayden O, Agarwal R, Lieber CM: Nanoscale avalanche photodiodes for highly sensitive and spatially resolved photon detection.*Nature Mater* 2006, 5:352-356.
4. Lin Wang Z: Zinc oxide nanostructures: growth, properties and applications.*J Phys Condens Matter* 2004, 16R:829-R858.

5. Cheng Liu W, Cai W: One dimensional and quasi-one-dimensional ZnO nanostructure prepared by spray pyrolysis-assisted thermal evaporation.*Appl Surf Sci* 2008, 254:3162-3166.
6. Wang EW, Sheehan PE, Lieber CM: Nanobeam mechanics: elasticity, strength and toughness of nanorods and nanotubes. *Science* 1971, 1997:277.
7. Harold S, Park : Stress-induced martensitic phase transformation in inter metallic nickel aluminum nanowires.*Nano Lett* 2006, 6:958-962.
8. Akram IB, Bunimovich Y, Jamil Tahir K, Jen-Kan Y, William GA, Heath JR: Silicon nanowires as efficient thermoelectric materials. *Nature* 2008, 451:168-171.
9. Hsieh Y-P, Chen H-Y, Lin M-Z, Shiu S-C, Hofmann M, Chern MY, Jia X, Yang YJ, Chang HJ, Huang HM, Tseng SC, Chen LC, Chen KH, Lin CF, Liang CT, Chen YF: Electroluminescence from ZnO/Si-nanotips light-emitting diodes.*Nano Lett* 2009, 9:5.1839-1843.
10. Wang JX, Sun XW, Wei A, Lei Y, Cai XP, Li CM, Dong ZL: Zinc oxide nanocomb biosensor for glucose detection.*Appl Phys Lett* 2006, 88:233106.
11. Wang JX, Sun XW, Yang Y, Huang H, Lee YC, Tan OK, Vayssieres L: Hydrothermally grown oriented ZnO nanorod arrays for gas sensing applications.*Nanotechnology* 2006, 17:4995-4998.
12. Cui Y, Lieber CM: Functional nanoscale electronic devices assembled using silicon nanowire building blocks.*Science* 2001, 291:851.
13. Javey A, Nam SW, Friedman RS, Yan H, Lieber CM: Layer-by-layer assembly of nanowires for three-dimensional, multifunctional electronics.*Nano Lett* 2007, 7:773.
14. Jiang BP, Zhou J-J, Fang H-F, Wang C-Y, Wang ZL, Si-S X: Hierarchical shelled ZnO structures made of bunched nanowire arrays.*Adv Funct Mater* 2007, 17:1303-1310.
15. Huang M, Mao S, Feick H, Yan H, Wu Y, Kind H, Weber E, Russo R, Yang P: Room-temperature ultraviolet nanowire nanolasers. *Science* 1897, 2001:292.
16. Sun XW, Kwok HS: Optical properties of epitaxially grown zinc oxide films on sapphire by pulsed laser deposition.*J Appl Phys* 1999, 86:408-411.

17. Lee CJ, Lee TJ, Lyu SC, Zhang Y, Ruh H, Lee H: Field emission from well-aligned zinc oxide nanowires grown at low temperature.*J Appl Phys Lett* 2002, 81:3648-3650.
18. Sun XW, Yu SF, Xu CX, Yue C, Chen BJ, Li S: Room-temperature ultraviolet lasing from zinc oxide microtubes.*Jpn J Appl Phys* 2003, 42:L1229-L1231.
19. Huang MH, Mao S, Feick H, Yan HQ, Wu YY, Kind H, Weber E, Russo R, Yang PD: Room-temperature ultraviolet nanowire nanolasers.*Science* 2001, 292:1897-1899.
20. Regan BO, Gratzel M: A low-cost, high-efficiency solar cell based on dye sensitized colloidal TiO_2 films.*Nature* 1991, 353:737-740.
21. Jiang CY, Sun XW, Lo GB, Kwong DL, Wang JX: Improved dye-sensitized solar cells with a ZnO-nanoflower photoanode.*Appl Phys Lett* 2007, 90:263501.
22. Wang ZL, Song JH: Piezoelectric nanogenerators based on zinc oxide nanowire arrays.*Science* 2006, 312:242.
23. Li Z, Yang RS, Yu M, Bai F, Li C, Wang ZL: Cellular level biocompatibility and biosafety of ZnO nanowires.*J Phys Chem C* 2009, 112:20114.
24. Harriott LR: Scattering with angular limitation projection electron beam lithography for suboptical lithography.*J Vac Sci Technol B* 1997, 15(6):2130-2135.
25. Chou Stephen Y, Peter KR, Preston RJ: Nanoimprint lithography.*J Vac Sci Technol B* 1996, 14(6):4129-4133.
26. Cheng C, Lei M, Feng L, Lun Wong T, Ho KM, Fung KK, Loy MM, Yu D, Wang N: High-quality ZnO nanowire arrays directly fabricated from photoresists.*ACS Nano* 2009, 3(1):53-58.
27. Ren S, Bai YF, Chen J, Deng SZ, Xu NS, Wu QB, Yang S: Catalyst-free synthesis of ZnO nanowire arrays on zinc substrate by low temperature thermal oxidation.*Mater Lett* 2007, 61:666.
28. Zhang Y, Jia HB, Wang RM, Chen CP, Luo XH, Yu D: Low-temperature growth and Raman scattering study of vertically aligned ZnO nanowires on Si substrate.*Appl Phys Lett* 2003, 83:4631.
29. Wilander M, Klason P, Yang LL, Safaa M, Al-Hilli SM, Zhao QX, Nur O: ZnO nanowires: chemical growth, electro-deposition and

application to intracellular nano-sensors.*Phy Stat Sol(c)* 2008, 5:3076-3083.

30. Vayssieres L: Growth of arrayed nanorods and nanowires of ZnO from aqueous solutions.*Adv Mater* 2003, 15:464.
31. Leprince-Wang Y, Yacoubi-Ouslim A, Wang GY: Structural study of electrodeposited ZnO nanowires.*Microelectro J* 2005, 36:625.
32. Azaceta E, Tena-Zaera R, Marcilla R, Fantini S, Echeberria J, Pomposo JA, Grande H, Mecerreyes D: Electrochemical deposition of ZnO in a room temperature ionic liquid: 1-butyl-1-methylpyrrolidinium bis(trifluoromethane sulfonyl)imide. *Electrochem Commun* 2009, 11:2184-2186.
33. Sheng X, Yong Q, Chen X, Yaguang W, Rusen Y, Zhong WL: Self-powered nanowire devices.*Nat Nanotechnol* 2010, 5:366-373.10.1038/NANO. 2010.46.
34. Ozen I, Ali Gulgun M: Residual stress relaxation and microstructure in ZnO thin films.*Adv Sci Technol* 2006, 45:1316.
35. Zhang Y, Du G, Liu D, Wang X, Ma Y, Wang J, Yin J, Yang X, Hou X, Yang S: Crystal growth of undoped ZnO films on Si substrates under different sputtering conditions.*J Crystal Growth* 2002, 243:439-443.
36. Seipel B, Nadarajah A, Wutzke B, Könenkamp R: Electrodeposition of ZnO nanorods in the presence of metal ions.*Mater Lett* 2009, 63:736-738.
37. Agrawal R, Peng B, Gdoutos EE, Espinosa HD: Elasticity size effects in ZnO nanowires – a combined experimental-computational approach.*Nano Lett* 2008, 11:3668-3674.
38. Cammarata RC: Surface and interface stress effect in thin films. *Prog Surf Sci* 1994, 46:11-38.
39. Lo S-S, Huang D, Tu C-H, Jan D-J: Formation and Raman scattering of seed-like ZnO nanostructure.*J Raman Spectrosc* 2009, 40:1694-1697.
40. Xu Q, Zhou S, Schmidt H: Magnetic properties of ZnO nanopowders.*J Alloys Compd* 2009, 487:665-667.
41. Sekaric L, Gunawan O, Majumdar A, Hu Liu X, Weinstein D, Sleight JW: Size-dependent modulation of carrier mobility in top-down fabricated silicon nanowires.*Appl Phys Lett* 2009, 95:023113.

42. Bouderbala M, Hamzaoui S, Adnane M, Sahraoui T, Zerdali M: Annealing effect on properties of transparent and conducting ZnO thin films.*Thin Solid Films* 2009, 517:1572-1576.
43. Sharma SK, Rammohan A, Sharma A: Templated one step electrodeposition of high aspect ratio n-type ZnO nanowire arrays.*J Colloid Interface Sci* 2010, 344:1-9.
44. Li Q, Wei L, Xie Y, Zhang K, Liu L, Zhu D, Jiao J, Chen Y, Yan S, Liu G, Mei L: ZnO nanoneedle/H_2O solid-liquid heterojunction-based self-powered ultraviolet detector.*Nanoscale Res Lett* 2013, 8:415.
45. Yoshida T, Komatsu D, Shimokawa N, Minoura H: Mechanism of cathodic electrodeposition of zinc oxide thin films from aqueous zinc nitrate baths.*Thin Solid Films* 2004, 166:166.
46. Dongping Z, Ping F, Xingmin C, Jianjun H, Lili R, Zhuangha Z, Guangxing L, Yukun H:Properties of ZnO thin films deposited by DC reactive magnetron sputtering under different plasma power. *Applied Physics A* 2009, 97:437-441.
47. Nainaparampil JJ, Zabinski JS, Prasad SV: Nanotribology of single crystal ZnO surfaces restructuring at high temperature annealing.*J Vac Sci Technol A* 1999, 17(4):1787-1792.
48. Puchert MK, Timbrell PY, Lamb RN: Postdeposition annealing of radio frequency magnetron sputtered ZnO films.*J Vac Sci Technol A* 1996, 14(4):2220.
49. Meng F, Yan X, Zhu Y, Si P: Controllable synthesis of MnO_2/polyaniline nanocomposite and its electrochemical capacitive property.*Nanoscale Res Lett* 2013, 8:179.
50. Fang ZB, Yan ZJ, Tan YS, Liu XQ, Wang YY: Influence of post-annealing treatment on the structure properties of ZnO films. *Appl Surf Sci* 2005, 241:303-308.
51. Tiwari A, Park M, Jin C, Eang H, Kumar D, Narayan J: Epitaxial growth of ZnO films on Si(111).*J Mater Res* 2002, 17:102,480-2483.
52. Kulkarni AJ, Zhou M, Kanokanan S, Limpijumnong S: Novel phase transformation in ZnO nanowires under tensile loading. *Phys Rev Lett* 2006, 97:105502.
53. Lee HW, Lau SP, Wang YG, Tay BK, Hng HH: Internal stress and surface morphology of zinc oxide thin films deposited by filtered cathodic vacuum arc technique.*Thin Solid films* 2004, 458:15.

54. Claeyssens F, Freeman CL, Allan NL, Sun Y, Ashfold MNR, Harding JH: Growth of ZnO thin films—experiment and theory.*J Mater Chem* 2005, 15:139-148.139

55. Sano M, Adaniya T, Fujitani T, Nakamura J: Oxidation of a Zn-deposited Cu(1 1 1) surface studied by XPS and STM.*Surf Sci* 2002, 514:261-266.

56. Özgür Ü, Alivov YI, Liu C, Teke A, Reshchikov MA, Do an S, Avrutin V, Cho S-J, Morkoç H: A comprehensive review of ZnO materials and devices.*J Appl Phys* 2005, 98:041301.

57. BenYahia S, Znaidi L, Kanaev A, Petitet JP: Raman study of oriented ZnO thin films deposited by sol–gel method.*Spectrochim Acta A* 2008, 711:234.

58. Alim KA, Fonoberov VA, Shamsa M, Balandin AA: Micro-Raman investigation of optical phonons in ZnO nanocrystals.*J Appl Phys* 2005, 593:124313.

59. Golan G, Axelevitch A, Gorenstein B, Maneych V: Hot-probe method for evaluation of impurities concentration in semiconductors.*Microelectronics J* 2006, 37:910.

60. Kishimoto S, Amamoto TY, Nakagawa Y, Lkeda K, Makino H, Yamada T: Dependence of electrical and structural properties on film thickness of undoped ZnO thin films prepared by plasma-assisted electron beam deposition.*Superlattice Microst* 2006, 39:306.

61. Hoa Hong N, Sakai J, Briz'e V: Observation of ferromagnetism at room temperature in ZnO thin films.*J Phys Condens Mat* 2007, 19:036219.

62. Hoa Hong N, Sakai J, Poirot N, Brizé V: Room-temperature ferromagnetism observed in undoped semiconducting and insulating oxide thin films.*Phys Rev B* 2006, 73:132404.

63. Khalid M, Ziese M, Setzer A, Esquinazi P, Lorenz M, Hochmuth H, Grundmann M, Spemann D, Butz T, Brauer G, Anwand W, Fischer G, Adeagbo WA, Hergert W, Ernst A: Defect-induced magnetic order in pure ZnO films.*Phys Rev B* 2009, 80:035331.

Citations

CHAPTER 1

Kai-Yu Peng and Da-Hua Wei, "Improving Light Extraction of Organic Light-Emitting Devices by Attaching Nanostructures with Self-Assembled Photonic Crystal Patterns," International Journal of Photoenergy, vol. 2014, Article ID 936049, 9 pages, 2014. doi:10.1155/2014/936049.

CHAPTER 2

T. Prakash, R. Jayaprakash, G. Neri, and Sanjay Kumar, "Synthesis of ZnO Nanostructures by Microwave Irradiation Using Albumen as a Template," Journal of Nanoparticles, vol. 2013, Article ID 274894, 8 pages, 2013. doi:10.1155/2013/274894.

CHAPTER 3

N. Gupta, G. F. Alapatt, R. Podila, R. Singh, and K. F. Poole, "Prospects of Nanostructure-Based Solar Cells for Manufacturing Future Generations of Photovoltaic Modules," International Journal of Photoenergy, vol. 2009, Article ID 154059, 13 pages, 2009. doi:10.1155/2009/154059.

CHAPTER 4

Heng Wu, Jun Zhou, Lizhi Liang, Lei Li, and Xinhua Zhu, "Fabrication, Characterization, Properties, and Applications of Low-Dimensional Bi-FeO_3 Nanostructures," Journal of Nanomaterials, vol. 2014, Article ID 471485, 14 pages, 2014. doi:10.1155/2014/471485.

CHAPTER 5

Lizhi Liang, Lei Li, Heng Wu, and Xinhua Zhu, Research Progress on Electronic Phase Separation in Low-Dimensional Perovskite Manganite Nanostructures, doi:10.1186/1556-276X-9-325.

CHAPTER 6

Wounjhang Park, Optical Interactions in Plasmonic Nanostructures, doi:10.1186/s40580-014-0002-x.

CHAPTER 7

Satinder K Shrama, Neelam Saurakhiya, Sumit Barthwal, Rudra Kumar, and Ashutosh Sharma, Tuning of Structural, Optical, and Magnetic Properties of Ultrathin and Thin ZnO Nanowire Arrays for Nano Device Applications, doi:10.1186/1556-276X-9-122.

Index